DOG ANATOMY COLORING BOOK

SCAN THE CODE TO ACCESS YOUR FREE DIGITAL COPY OF THE VETERINARY ANATOMY COLORING BOOK

SCAN ME

THIS BOOK BELONGS TO

TABLE OF CONTENTS

SECTION 1: THE SKELETON OF THE DOG LATERAL ASPECT

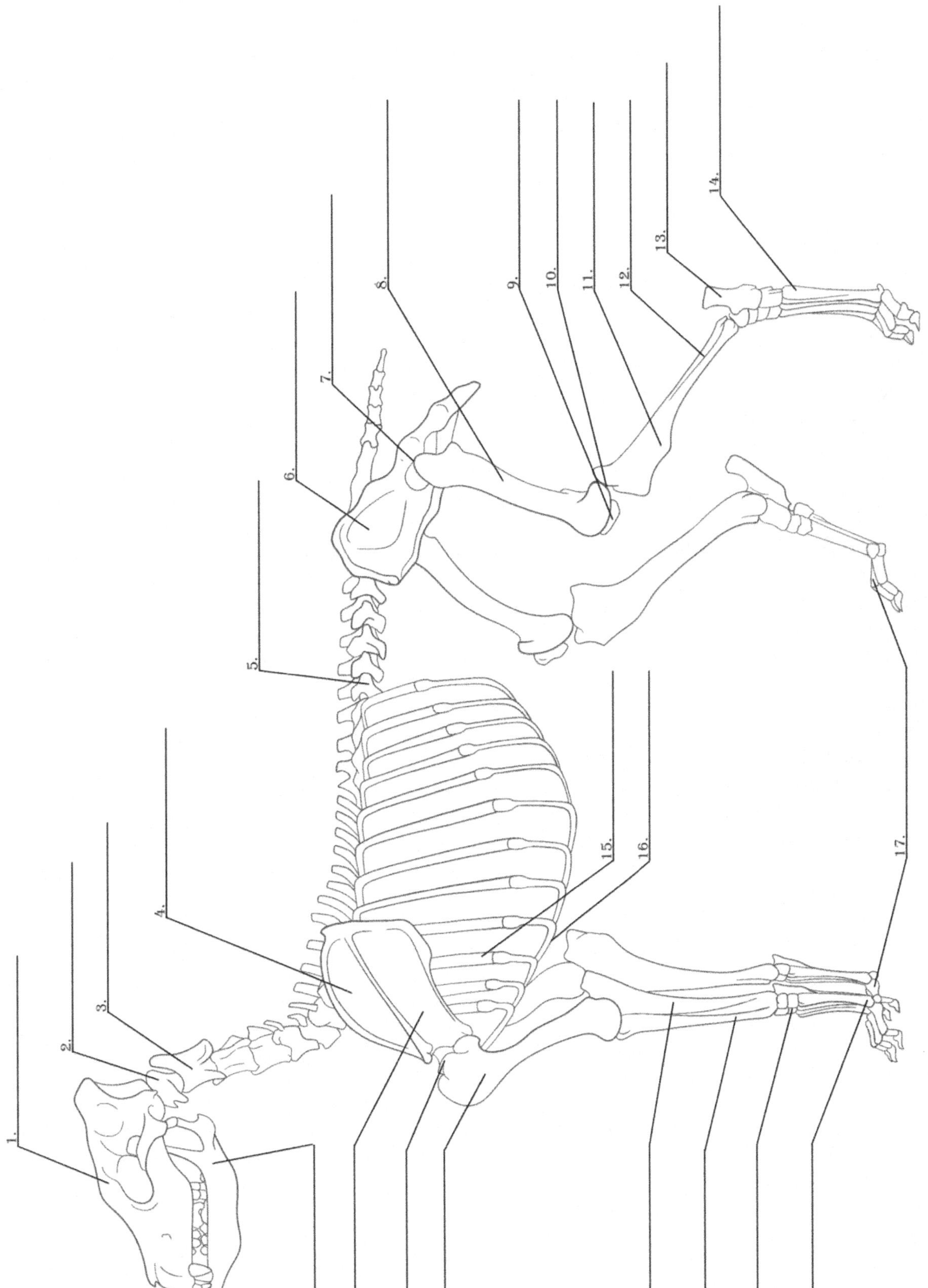

SECTION 1: THE SKELETON OF THE DOG LATERAL ASPECT

1.SKULL
2.ATLAS
3.AXIS
4.SCAPULA
5.SACRUM
6.PELVIS
7.HIP JOINT
8.FEMUR
9.PATELLA
10.STIFLE JOINT
11.TIBIA
12.FIBULA
13.HOCK JOINT
14.METATARSAL BONES
15.RIB
16.STERNUM
17.PHALANGES (TOE BONES)
18.MANDIBLE
19.SCAPULA
20.SHOULDER JOINT
21.HUMERUS
22.ULNA
23.RADIUS
24.CARPAL JOINT
25.METACARPAL BONES

SECTION 2:THE SKELETON OF THE DOG CRANIAL AND CAUDAL ASPECT

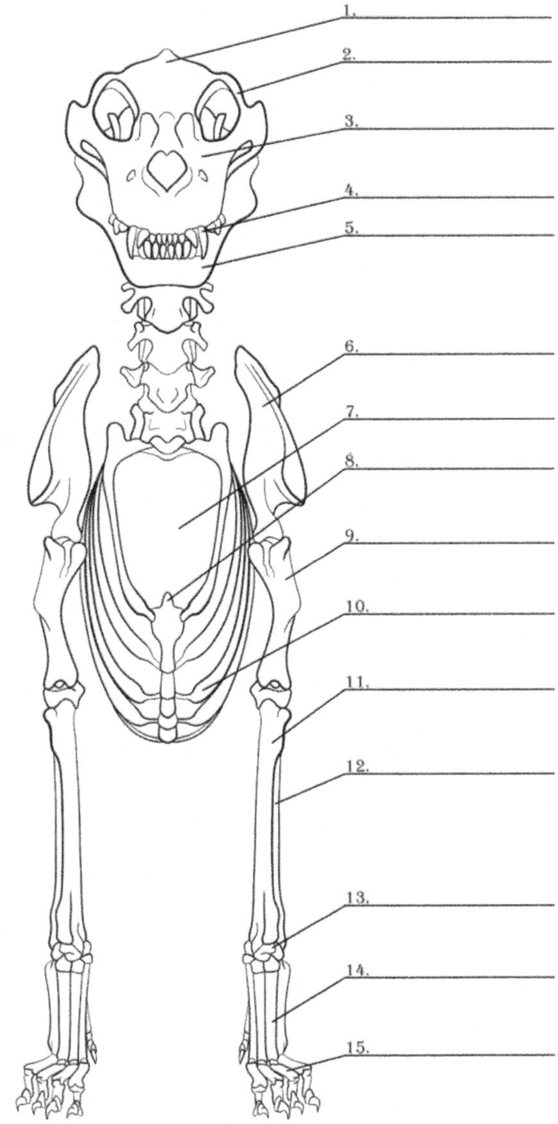

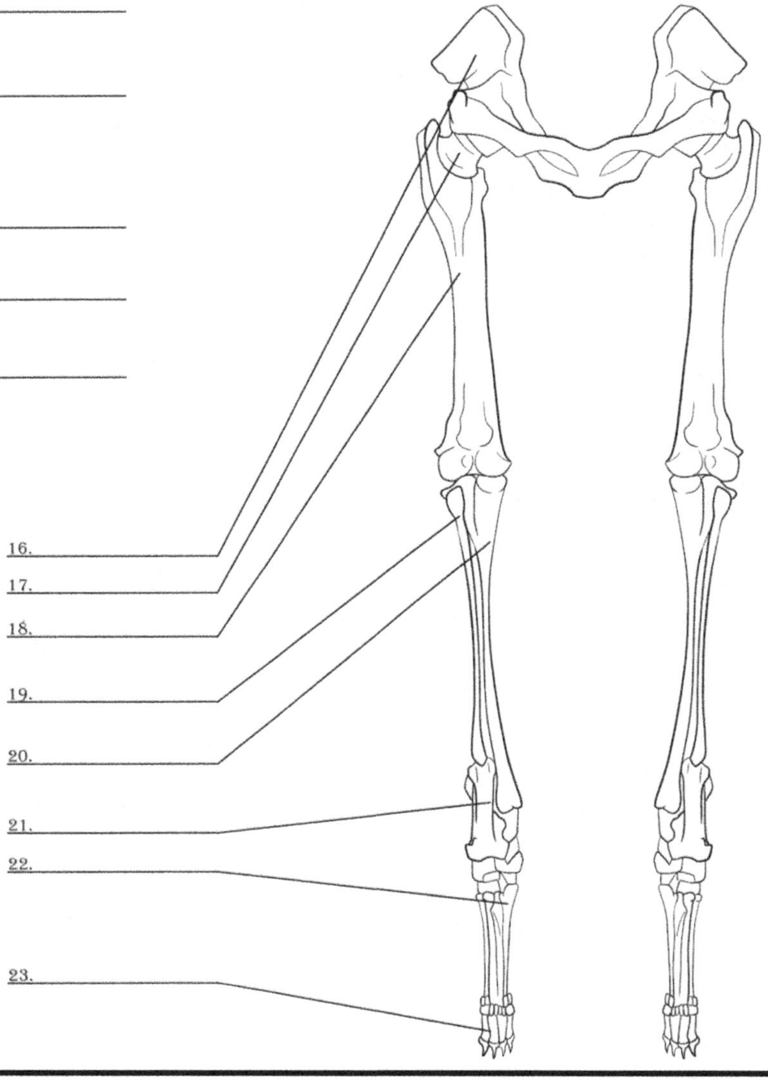

SECTION 2: THE SKELETON OF THE DOG CRANIAL AND CAUDAL ASPECT

1. OCCIPUT
2. SKULL
3. MAXILLA
4. TEETH
5. MANDIBLE
6. SCAPULA
7. BREAST CAVITY
8. STERNUM
9. HUMERUS
10. RIB
11. RADIUS
12. ULNA
13. CARPUS
14. METACARPUS
15. PHALANGES
16. PELVIS
17. HIP JOINT
18. FEMUR
19. FIBULA
20. TIBIA
21. HOCK JOINT
22. METATARSAL BONE
23. PHALANGES

SECTION 3: THE SKELETON OF THE DOG DORSAL ASPECT

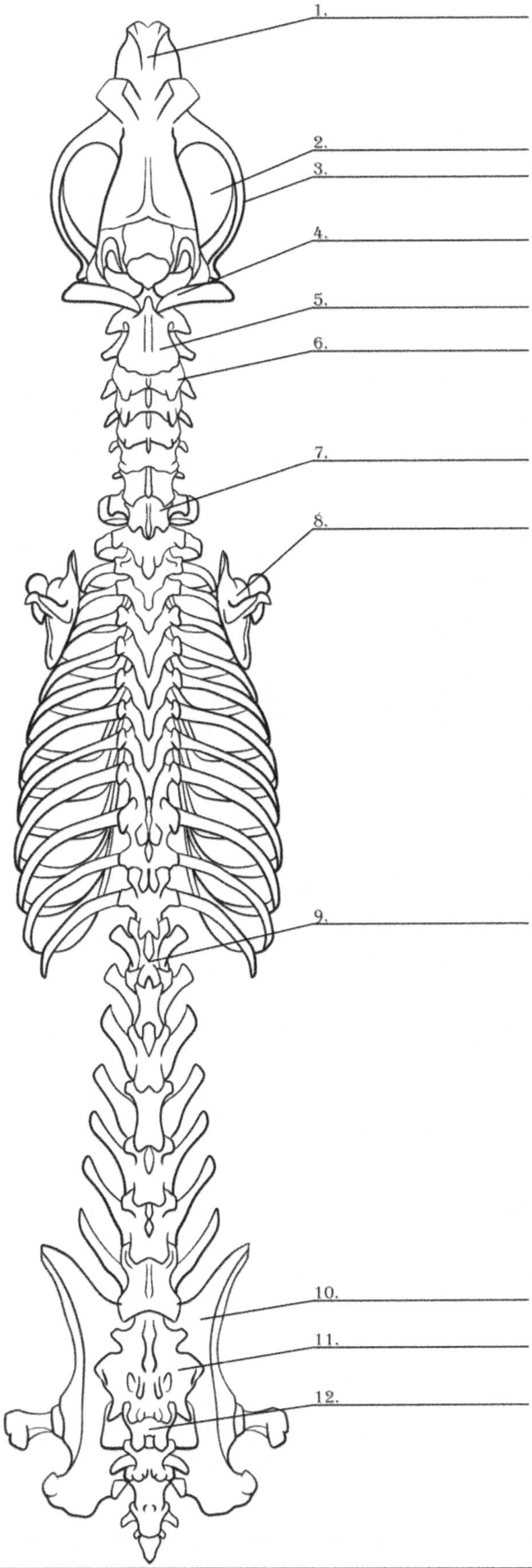

SECTION 3: THE SKELETON OF THE DOG DORSAL ASPECT

1. NASAL BONE
2. ORBIT
3. ZYGOMATIC ARCH
4. ATLAS
5. AXIS
6. CERVICAL VERTEBRAE
7. THORACIC VERTEBRAE
8. SCAPULA
9. LUMBAR VERTEBRAE
10. PELVIS
11. SACRUM
12. CAUDAL VERTEBRAE

SECTION 4: THE MUSCLES OF THE DOG LATERAL ASPECT

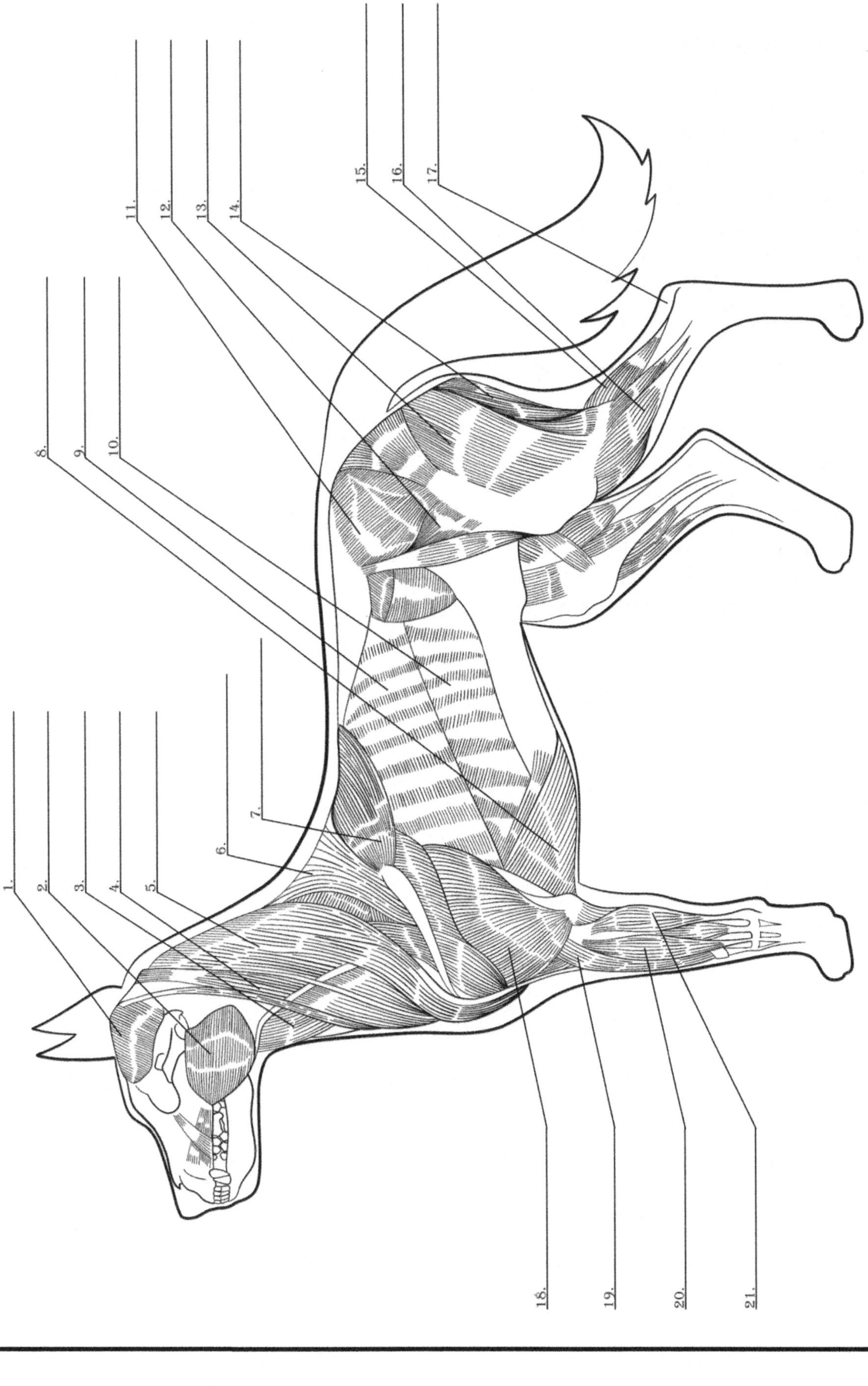

SECTION 4: THE MUSCLES OF THE DOG LATERAL ASPECT

1. TEMPORALIS MUSCLE
2. MASSETER MUSCLE
3. STERNOHYOID MUSCLE
4. STERNOCEPHALICUS MUSCLE
5. BRACHIOCEPHALICUS MUSCLE
6. TRAPEZIUS MUSCLE
7. DELTOID MUSCLE
8. DEEP PECTORAL MUSCLE
9. LATISSIMUS DORSI MUSCLE
10. EXTERNAL ABDOMINAL OBLIQUE MUSCLE
11. GLUTEAL MUSCLE
12. TENSOR FASCIAE LATAE MUSCLE
13. BICEPS FEMORIS MUSCLE
14. SEMITENDINOSUS MUSCLE
15. GASTROCNEMIUS MUSCLE
16. CRANIAL TIBIAL MUSCLE
17. ACHILLES TENDON
18. TRICEPS BRACHII MUSCLE
19. EXTENSOR CARPI RADIALIS MUSCLE
20. EXTENSOR CARPI ULNARIS MUSCLE
21. FLEXOR CARPI ULNARIS MUSCLE

SECTION 5: THE MUSCLES OF THE DOG CRANIAL AND CAUDAL ASPECT

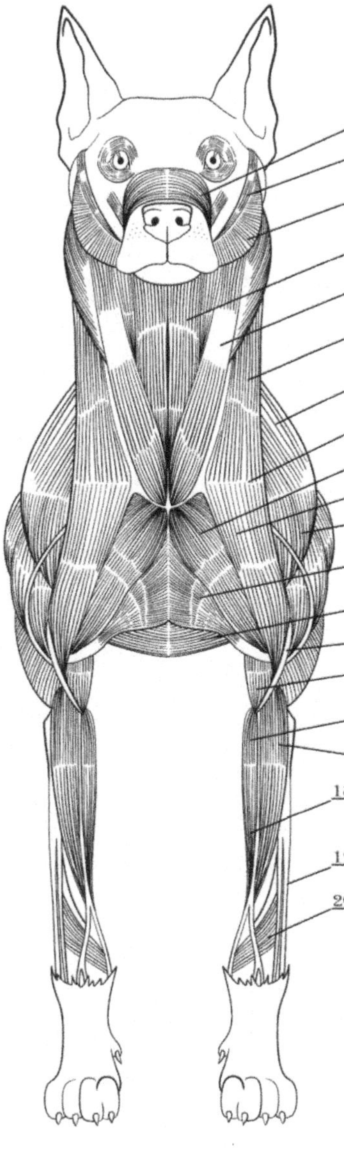

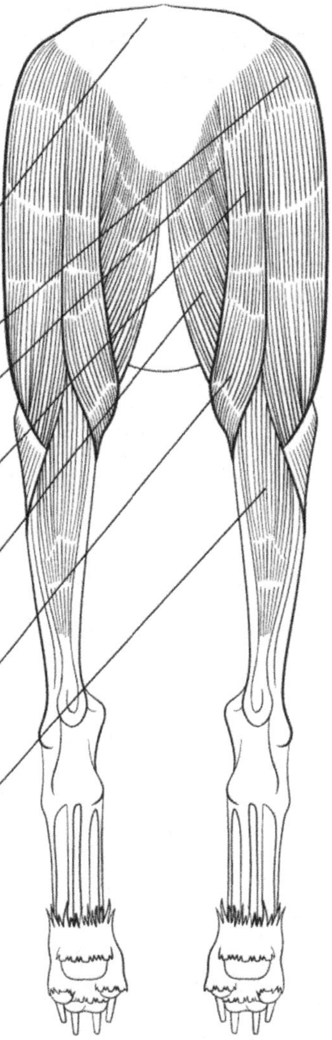

SECTION 5: THE MUSCLES OF THE DOG CRANIAL AND CAUDAL ASPECT

1. 1. NASOLABIAL LEVATOR MUSCLE
2. ZYGOMATIC MUSCLE
3. MASSETER MUSCLE
4. STERNOHYOID MUSCLE
5. STERNOCEPHALICUS MUSCLE
6. CLEIDOCEPHALICUS MUSCLE
7. OMOTRANSVERSARIUS MUSCLE
8. CLAVICULAR INTERSECTION
9. PECTORALIS DESCENDENS MUSCLE
10. CLEIDOBRACHIALIS MUSCLE
11. DELTOID MUSCLE
12. PECTORALIS SUPERFICIALIS MUSCLE
13. EXTERNAL ABDOMINAL OBLIQUE MUSCLE
14. BRACHIALIS MUSCLE
15. BICEPS BRACHII MUSCLE
16. PRONATOR TERES MUSCLE
17. EXTENSOR CARPI RADIALIS MUSCLE
18. FLEXOR CARPI RADIALIS MUSCLE
19. EXTENSOR DIGITORUM COMMUNIS MUSCLE
20. ABDUCTOR DIGITI MUSCLE

SECTION 6: THE MUSCLES OF THE DOG VENTRAL ASPECT

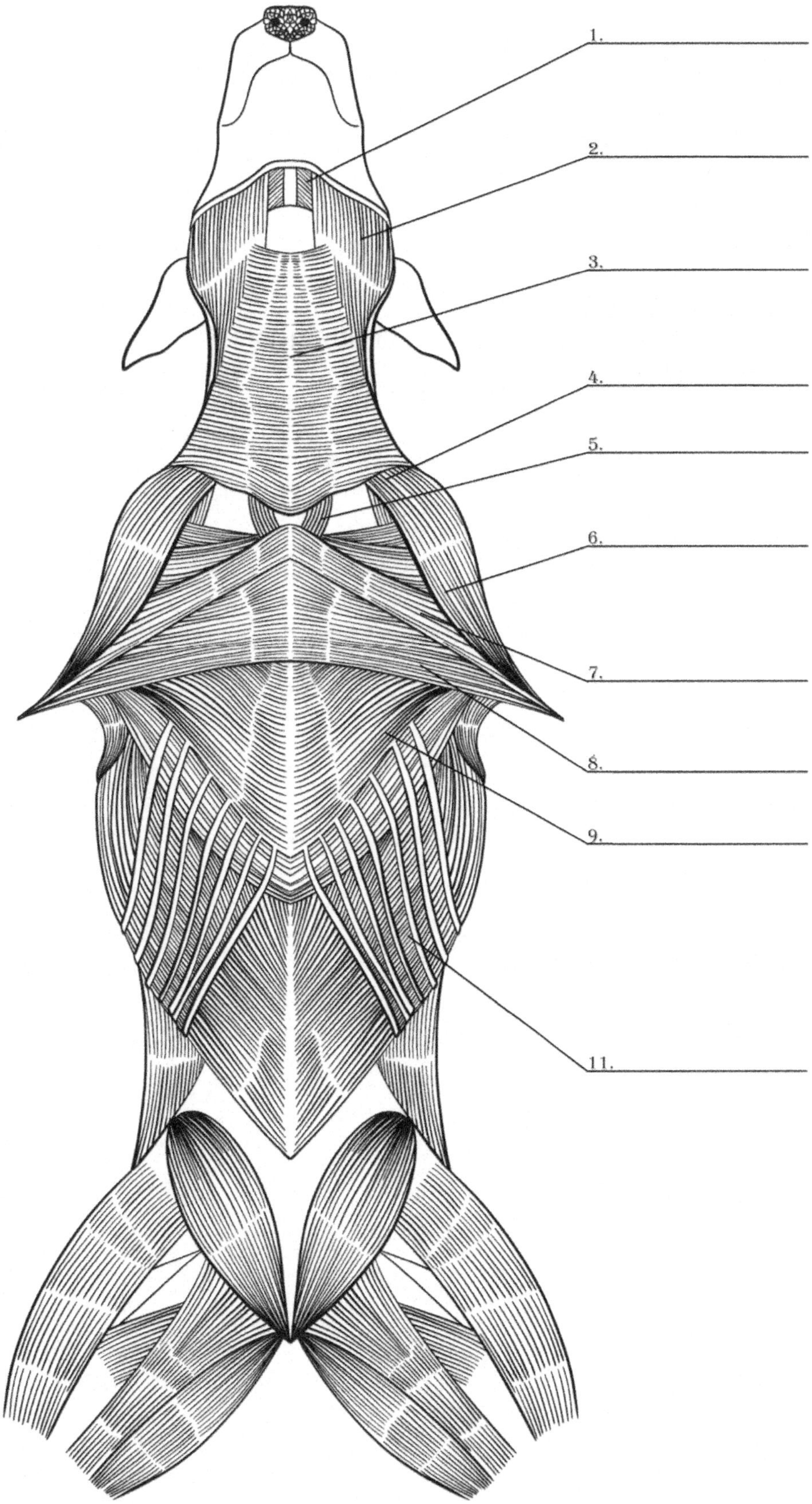

SECTION 6: THE MUSCLES OF THE DOG VENTRAL ASPECT

1. MUSCULUS MYLOHYOIDEUS
2. SPHINCTER COLLI PROFUNDUS MUSCLE
3. PLATYSMA MUSCLE
4. SPHINCTER COLLI SUPERFICIALIS MUSCLE
5. CLEIDOCEPHALICUS MUSCLE
6. STERNOCEPHALICUS MUSCLE
7. CLEIDOBRACHIALIS MUSCLE
8. PECTORALIS DESCENDENS MUSCLE
9. PECTORALIS TRANSVERSUS MUSCLE
10. PECTORALIS ASCENDENTE SUPERFICIALIS PROFUNDUS MUSCLE
11. CUTANEUS TRUNCI MUSCLE

SECTION 7: THE MUSCLES OF THE DOG DORSAL ASPECT

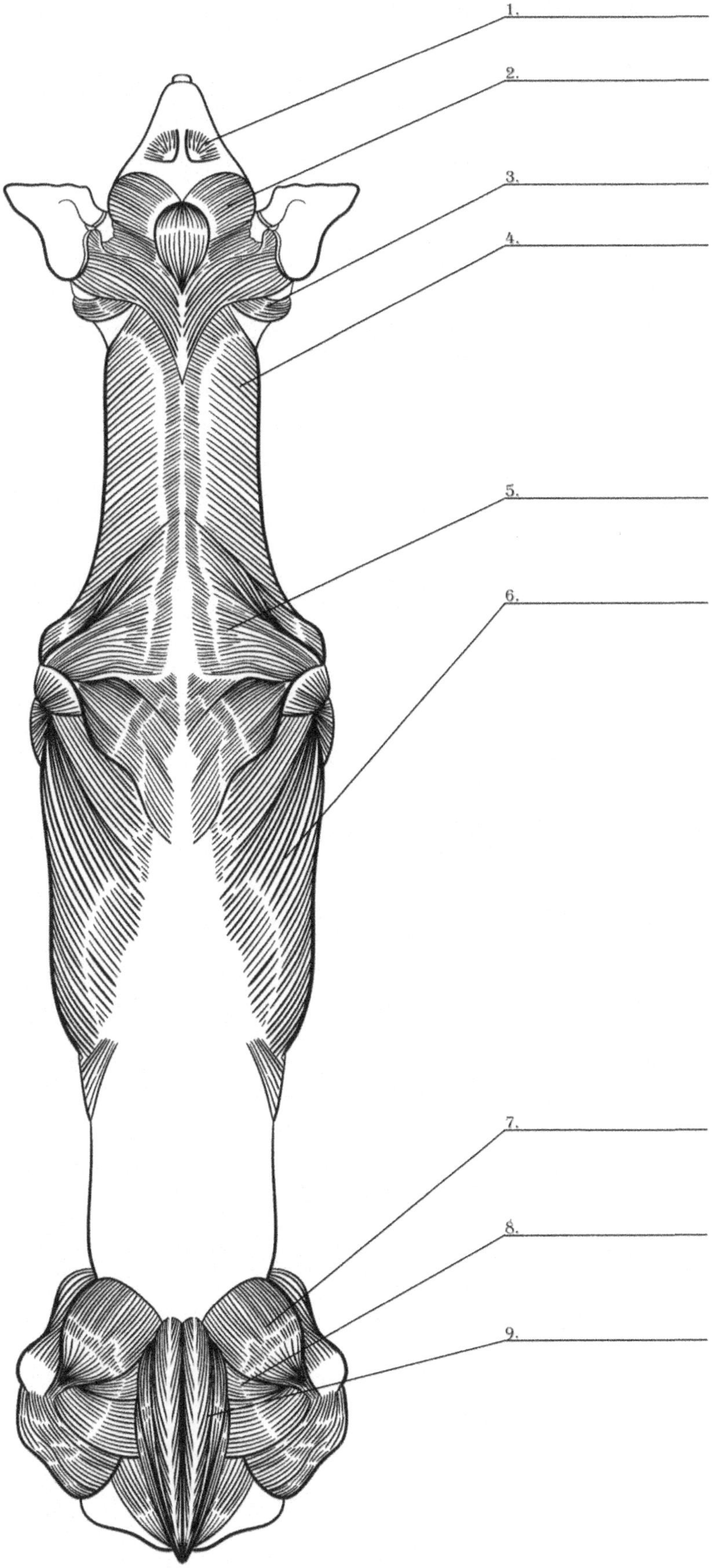

SECTION 7: THE MUSCLES OF THE DOG DORSAL ASPECT

1. LEVATOR NASOLABIALIS MUSCLE
2. PARS PALPEBRALIS MUSCLE
3. STERNOCEPHALICUS MUSCLE
4. CLEIDOBRACHIALIS MUSCLE
5. TRAPEZIUS MUSCLE
6. LATISSIMUS DORSI MUSCLE
7. GLUTEUS MEDIUS MUSCLE
8. GLUTEUS MAXIMUS MUSCLE
9. COCCYGEUS MUSCLE

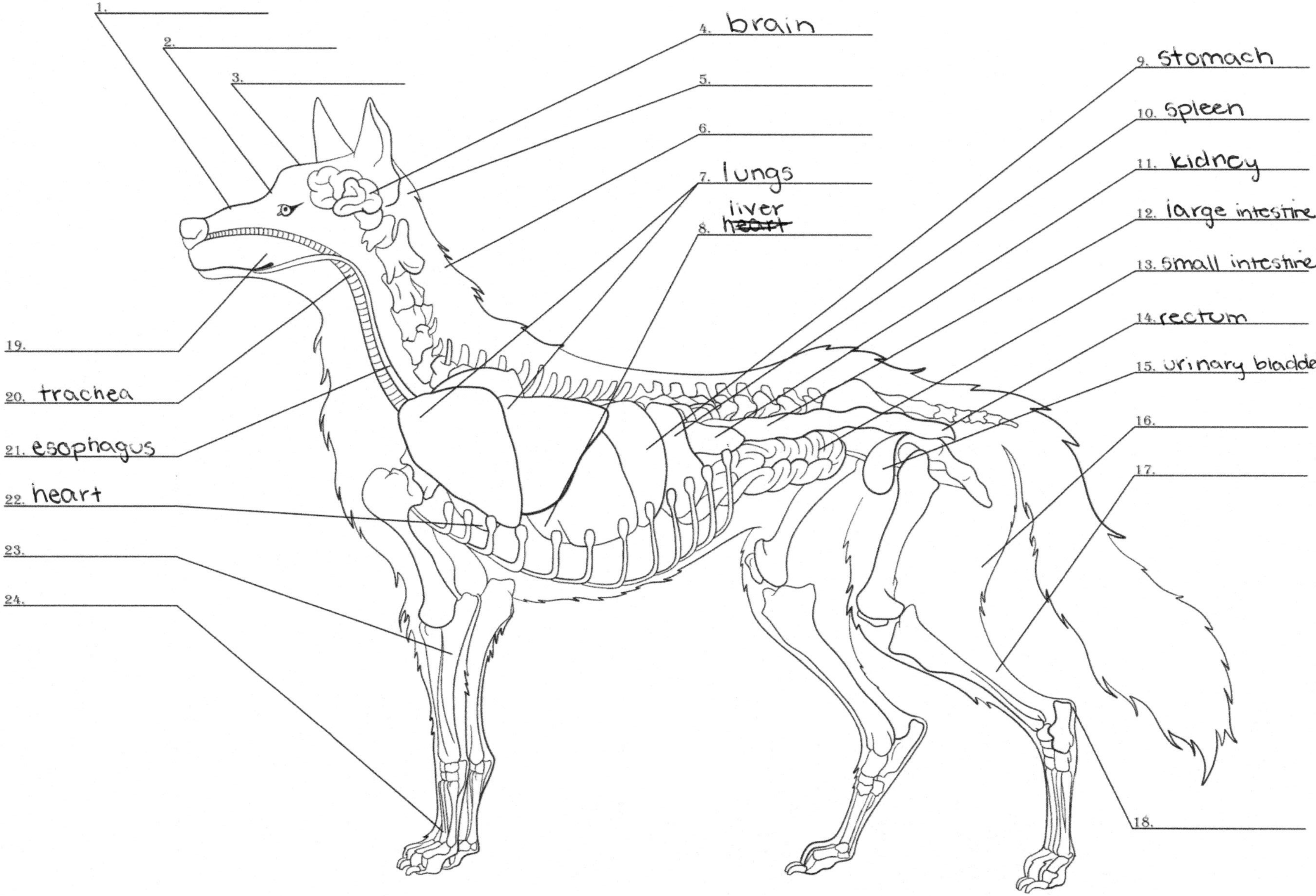
SECTION 8: INTERNAL ORGANS OF THE DOG
1.
2.
3.
4. brain
5.
6.
7. lungs
8. liver
9. stomach
10. spleen
11. kidney
12. large intestine
13. small intestine
14. rectum
15. urinary bladder
16.
17.
18.
19.
20. trachea
21. esophagus
22. heart
23.
24.

SECTION 8: INTERNAL ORGANS OF THE DOG

1. BRIDGE OF THE NOSE
2. STOP
3. UPPER SKULL
4. BRAIN
5. NAPE
6. NECK
7. LUNGS
8. LIVER
9. STOMACH
10. SPLEEN
11. KIDNEY
12. COLON
13. SMALL INTESTINE
14. RECTUM
15. BLADDER
16. UPPER TIGHT
17. LOWER TIGHT
18. POINT OF HOCK
19. MUZZLE
20. LARYNX
21. ESOPHAGUS
22. HEART
23. FOREARM
24. PASTERN

SECTION 9: BLOOD VESSELS OF THE DOG

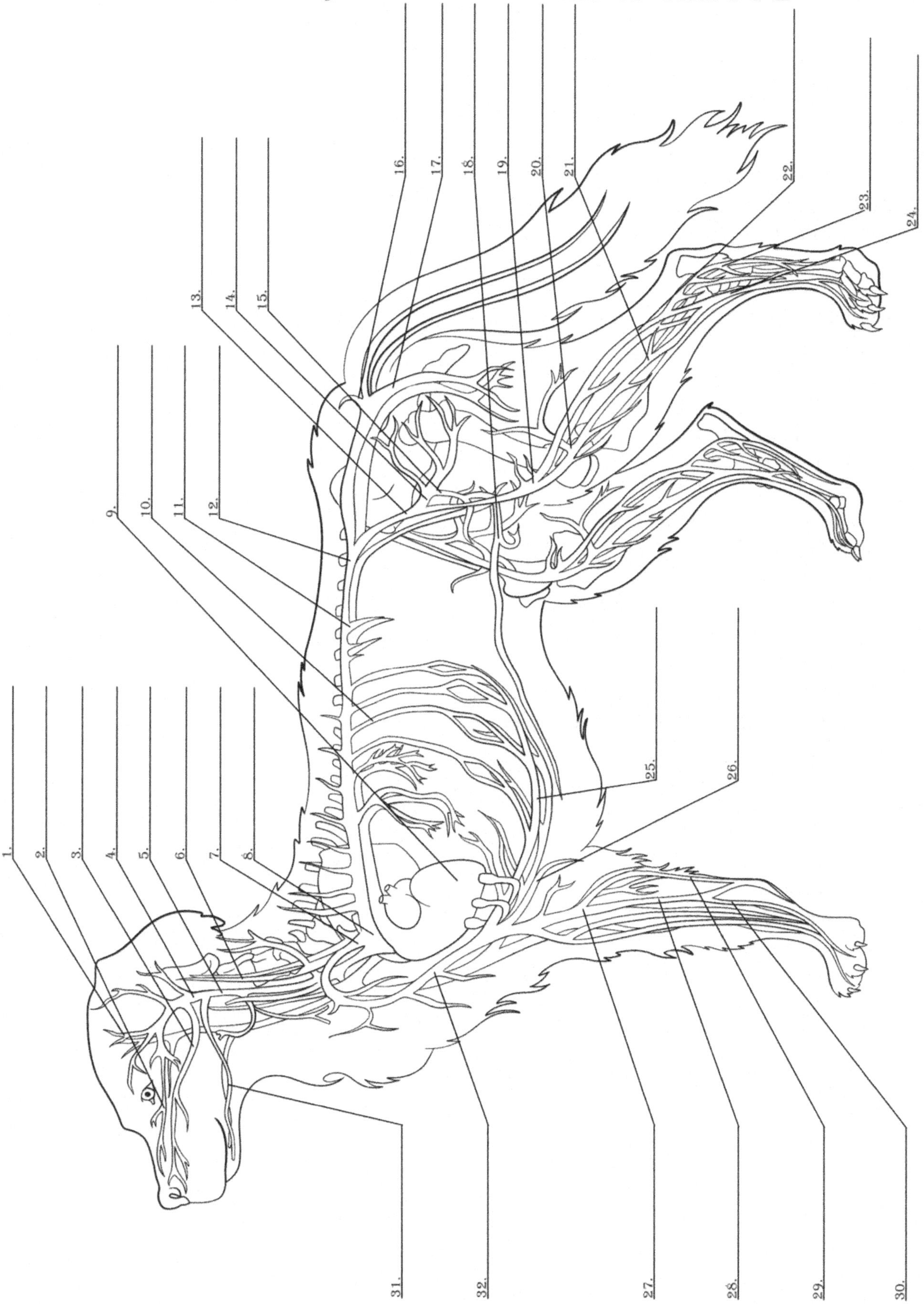

SECTION 9: BLOOD VESSELS OF THE DOG

1. SUPERFICIAL TEMPORAL ARTERY
2. INFRAORBITAL ARTERY
3. FACIAL ARTERY
4. INTERNAL CAROTID ARTERY
5. COMMON CAROTID ARTERY
6. VERTEBRAL ARTERY
7. LEFT SUBCLAVIAN ARTERY
8. AORTA
9. HEART
10. INTERCOSTAL ARTERY
11. RENAL ARTERY
12. ABDOMINAL AORTA
13. LEFT EXTERNAL ILIAC ARTERY
14. DEEP FEMORAL ARTERY
15. PUDENDOEPIGASTRIC TRUNK
16. CRANIAL GLUTEAL ARTERY
17. CAUDILLA GLUTEAL ARTERY
18. EXTERNAL PUDENDAL ARTERY
19. FEMORAL ARTERY
20. DISTAL CAUDAL FEMORAL ARTERY
21. CRANIAL TRIBAL ARTERY
22. SAPHENOUS ARTERY
23. CAUDAL BRANCH OF SAPHENOUS ARTERY
24. CRANIAL BRANCH OF SAPHENOUS ARTERY
25. INTERNAL THORACIC ARTERY
26. COLLATERAL ULNAR ARTERY
27. COMMON INTEROSSEOUS ARTERY
28. MEDIAN ARTERY
29. ULNAR ARTERY
30. RADIAL ARTERY
31. LINGUAL ARTERY
32. BRACHIAL ARTERY

SECTION 10: NERVES OF THE DOG

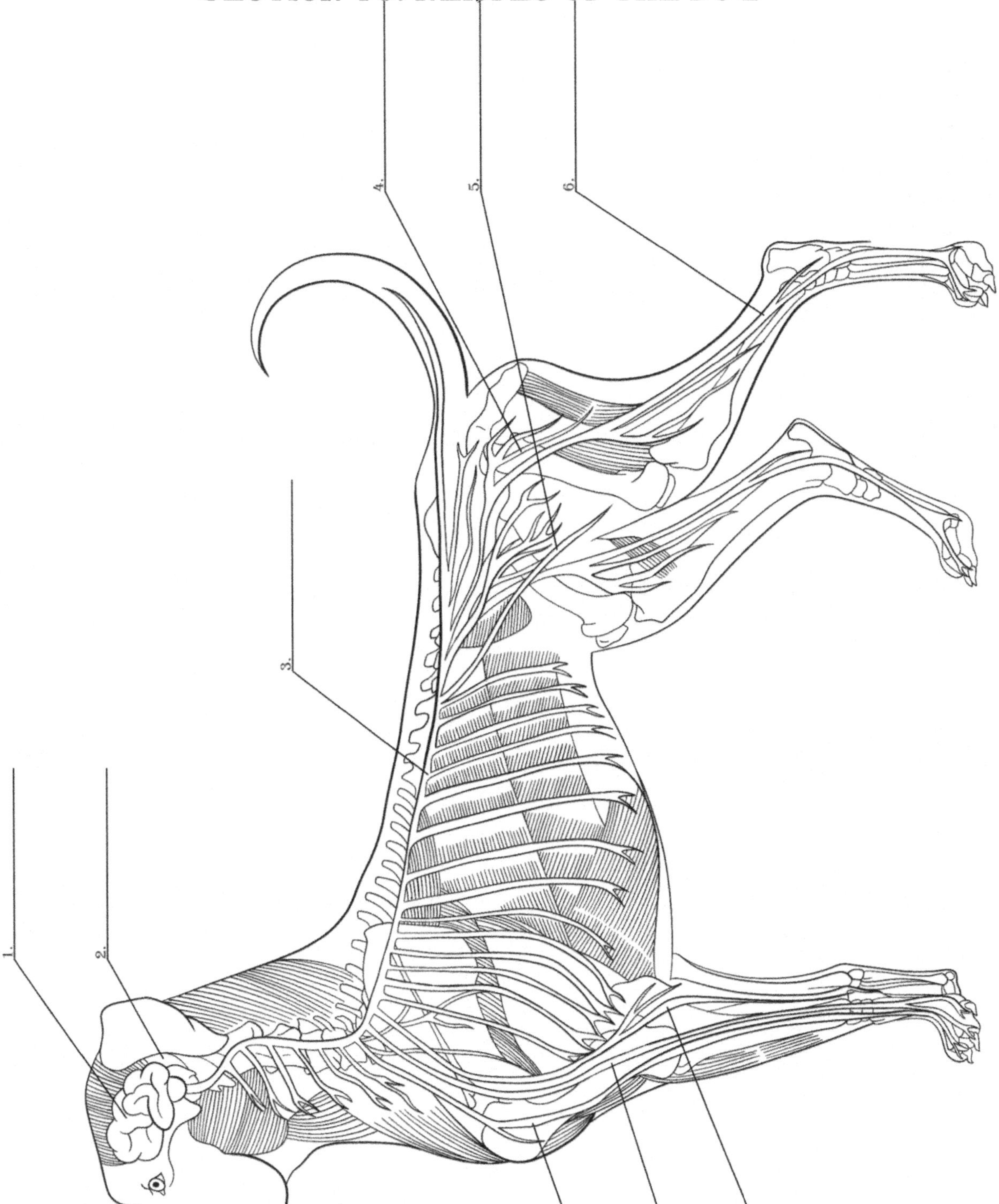

SECTION 10: NERVES OF THE DOG

1. CEREBRAL HEMISPHERE
2. CEREBELLUM
3. SPINAL CORD
4. SCIATIC NERVE
5. FEMORAL NERVE
6. TIBIAL NERVE
7. RADIAL NERVE
8. MEDIAL NERVE
9. ULNAR NERVE

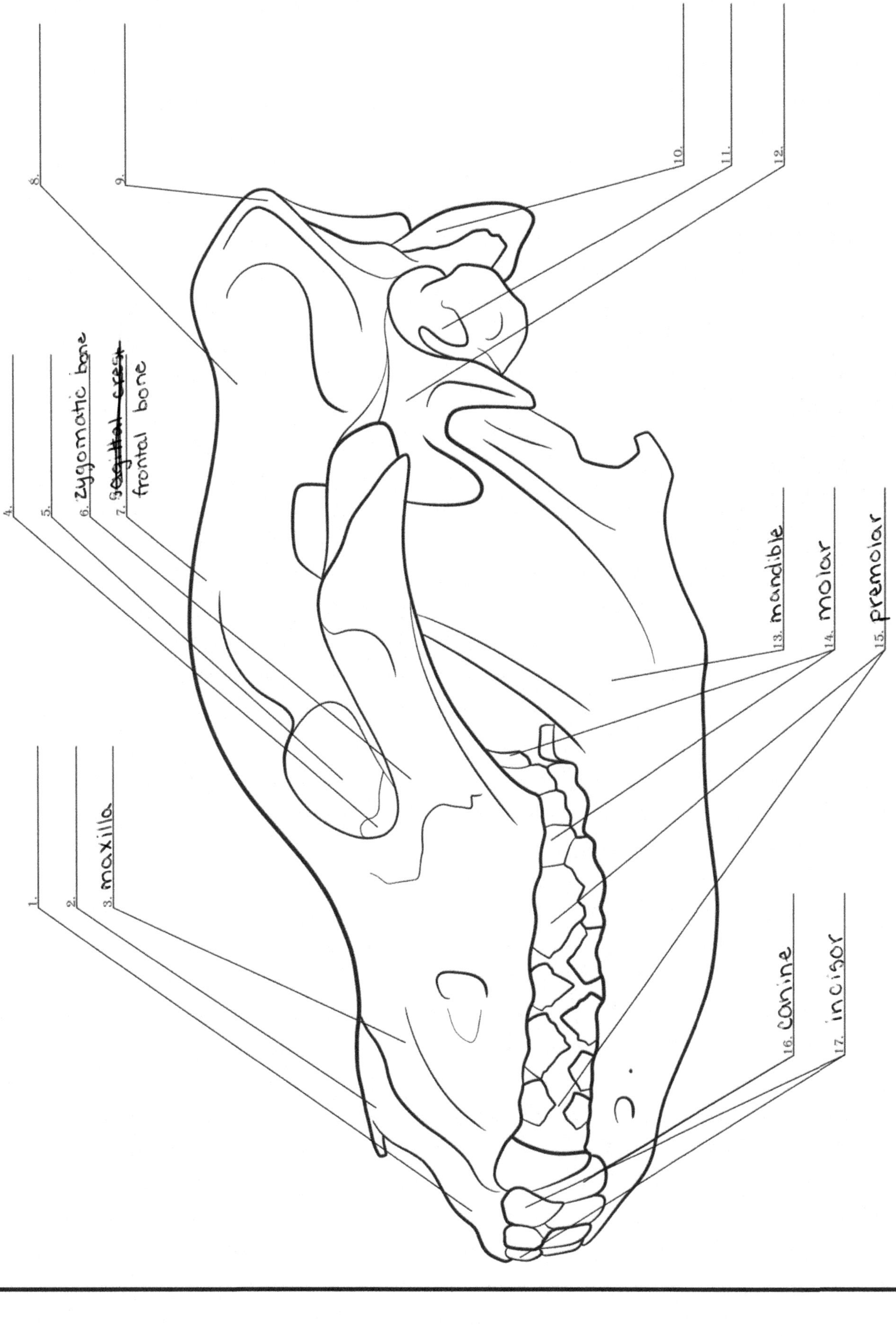
SECTION 11: THE SKULL OF THE DOG LATERAL ASPECT
1.
2.
3. maxilla
4.
5.
6. zygomatic bone
7. frontal bone
8.
9.
10.
11.
12.
13. mandible
14. molar
15. premolar
16. canine
17. incisor

SECTION 11: THE SKULL OF THE DOG LATERAL ASPECT

1. INCISIVE BONE
2. NASAL BONE
3. MAXILLA
4. LACRIMAL BONE
5. ORBIT
6. ZYGOMTIC BONE
7. FRONTAL BONE
8. PARIETAL BONE
9. OCCIPITAL BONE
10. OCCIPITAL CONDYLES
11. EXTERNAL AUDITORY MEATUS
12. TEMPORAL BONE
13. MANDIBLE
14. MOLAR TEETH
15. PREMOLAR TEETH
16. CANINE TEETH
17. INCISOR TEETH

SECTION 12: INSIDE THE SKULL OF THE DOG LATERAL ASPECT

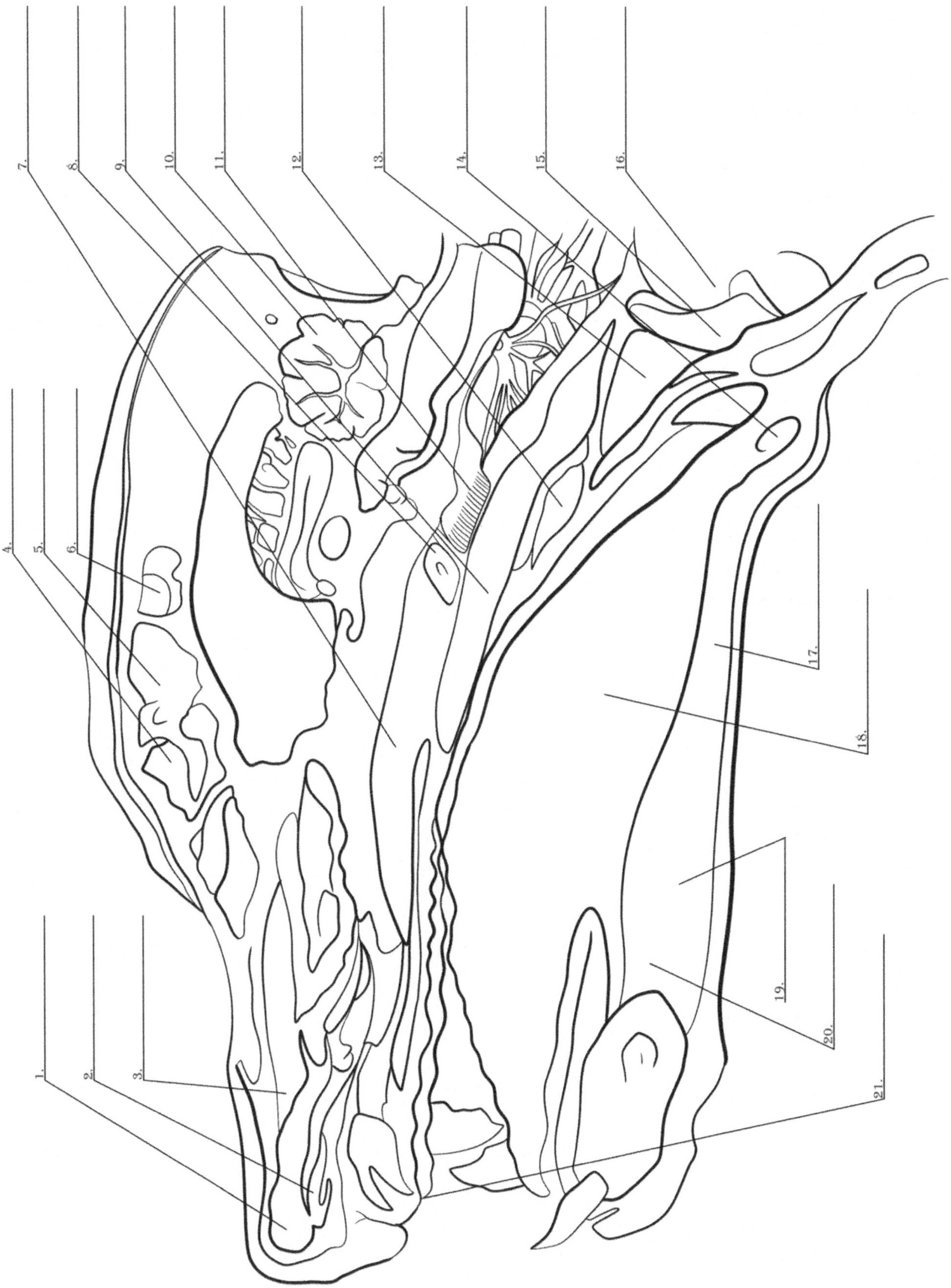

SECTION 12: INSIDE THE SKULL OF THE DOG LATERAL ASPECT

1. NASAL VESTIBULE
2. BASAL FOLD
3. STRAIGHTS FOLD
4. ROSTRAL FRONTAL SINUS
5. MEDIAL FRONTAL SINUS
6. LATERAL FRONTAL SINUS
7. PARS NASALIS
8. PHARYNGEAL OSTIUM OF AUDITORY TUBE
9. SOFT PALATE
10. CEREBELLUM
11. LEVATOR VELI PALATINI MUSCLE
12. PALATINE TONSIL
13. VESTIBULE OF LARYNX
14. BASIHYOID
15. VESTIBULAR FOLD
16. GLOTTIS
17. MYLOHYOID MUSCLE
18. LINGUALIS PROPRIUS MUSCLE
19. GENIOHYOID MUSCLE
20. GENIOGLOSSUS MUSCLE
21. VESTIBULE OF MOUTH

SECTION 13: THE SKULL OF THE DOG DORSAL ASPECT

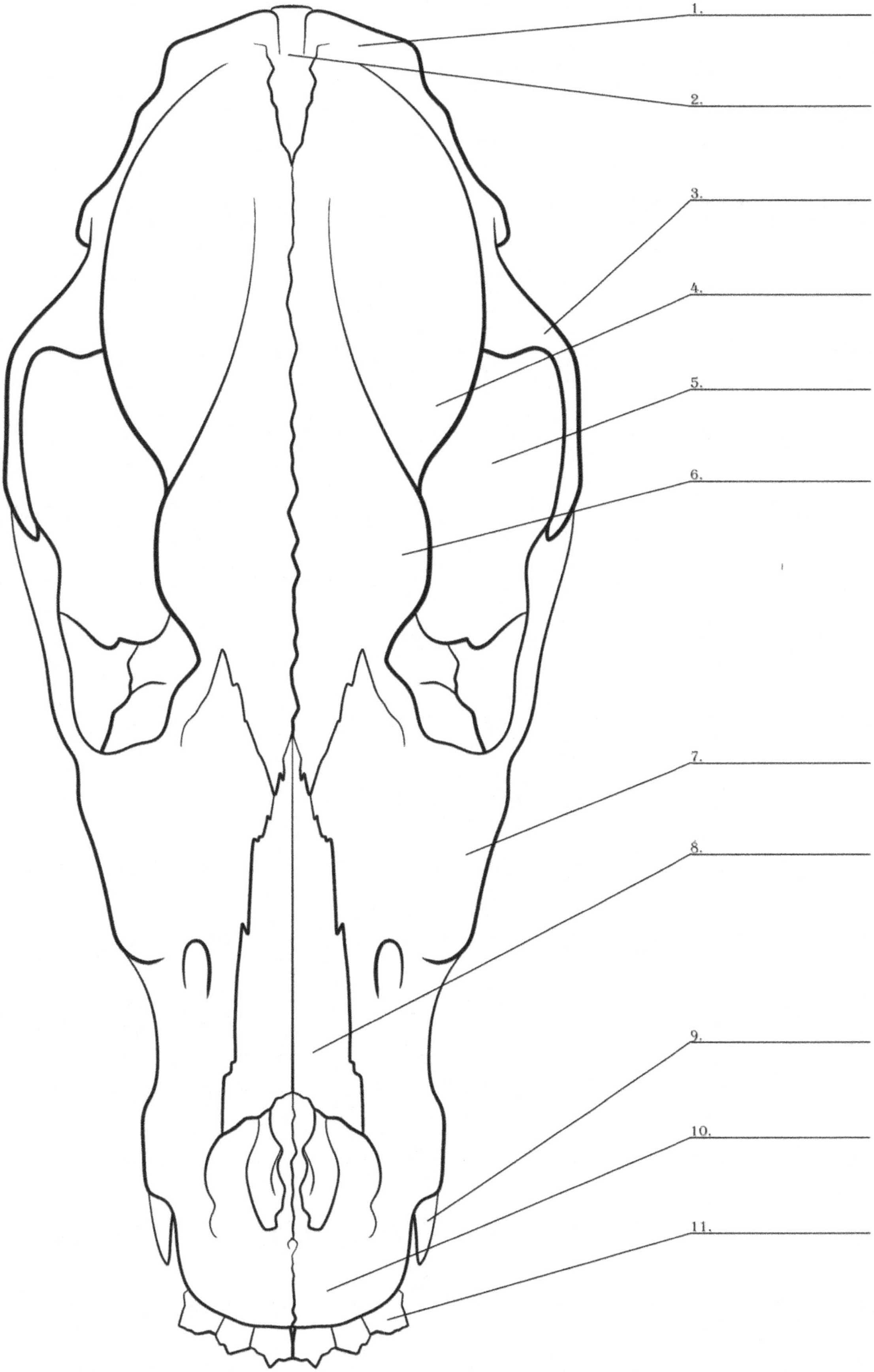

SECTION 13: THE SKULL OF THE DOG DORSAL ASPECT

1. NUCHAL CREST
2. MEDIAN OSSEOUS CREST
3. ZYGOMATIC ARCH
4. TEMPORAL FOSSA
5. ORBIT
6. ZYGOMATIC PROCESS OF THE FRONTAL BONE
7. FACIAL CREST
8. NASAL BONE
9. CANINE TEETH
10. INCISIVE BONE
11. INCISOR TEETH

SECTION 14: THE SKULL OF THE DOG VENTRAL ASPECT

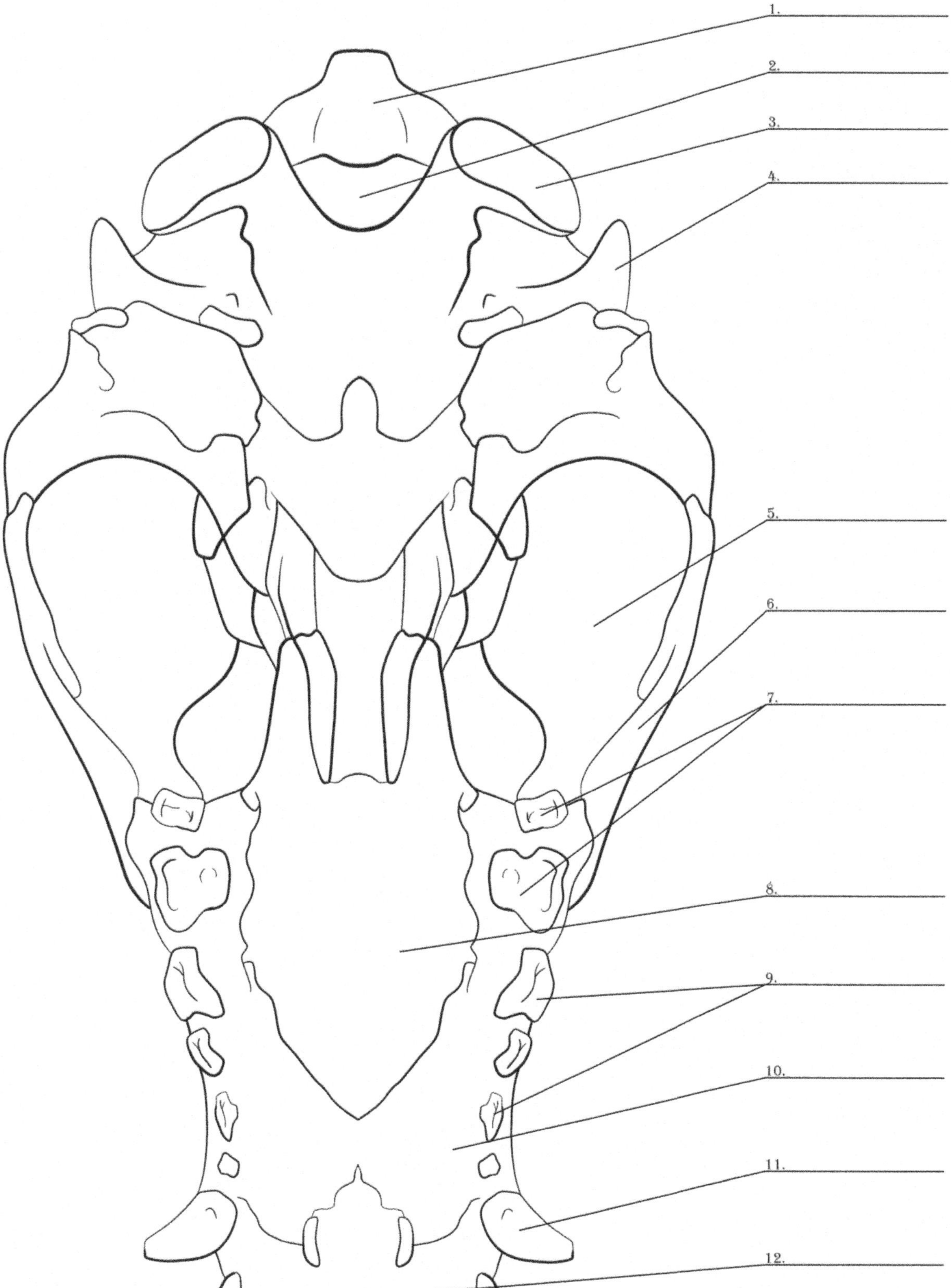

SECTION 14: THE SKULL OF THE DOG VENTRAL ASPECT

1. OCCIPITAL BONE
2. FORAMEN MAGNUM
3. OCCIPITAL CONDYLE
4. JUGULAR PROCESS
5. ORBIT
6. ZYGOMATIC ARCH
7. MOLAR TEETH
8. PALATINE BONE
9. PREMOLAR TEETH
10. MAXILLA
11. CANINE TEETH
12. INCISIVE BONE
13. INCISOR TEETH

SECTION 15: THE MUSCLES OF THE HEAD LATERAL ASPECT

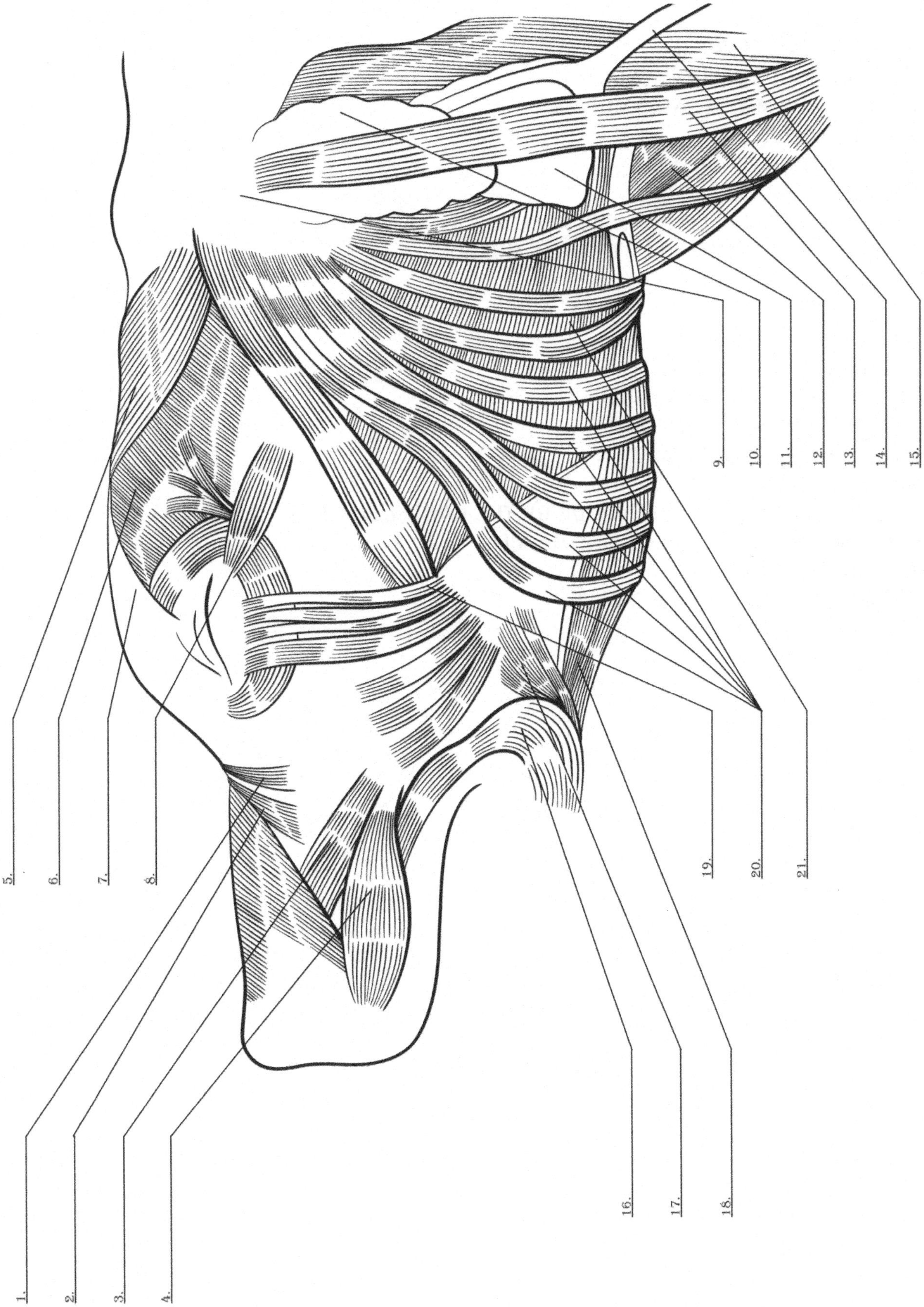

SECTION 15: THE MUSCLES OF THE HEAD LATERAL ASPECT

1. LATERALIS NASI MUSCLE
2. LEVATOR NASOLABIALIS MUSCLE
3. LEVATOR LABII MAXILLARIS MUSCLE
4. CANINUS MUSCLE
5. FRONTOSCUTULARIS MUSCLE
6. TEMPORALIS MUSCLE
7. LEVATOR ANGULI OCULI MEDIALIS MUSCLE
8. RETRACTOR ANGULI OCULI LATERALIS MUSCLE
9. SCUTIFORM CARTILAGE
10. PAROTID GLAND
11. MANDIBULAR GLAND
12. STERNOHYOIDEUS MUSCLE
13. PAROTIDEO-AURICULARIS MUSCLE
14. JUGULAR VEIN AND GROOVE
15. STERNOCEPHALICUS MUSCLE
16. ORBICULARIS ORIS MUSCLE
17. ZYGOMATICUS MUSCLE (ELEVATOR OF THE LABIAL ANGLE)
18. DEPRESSOR LABII MANDIBULARIS MUSCLE
19. MALARIA MUSCLE
20. ZYGOMATIC CUTANEOUS MUSCLES
21. MASSETER MUSCLE

SECTION 16: THE MUSCLES OF THE HEAD DORSAL ASPECT

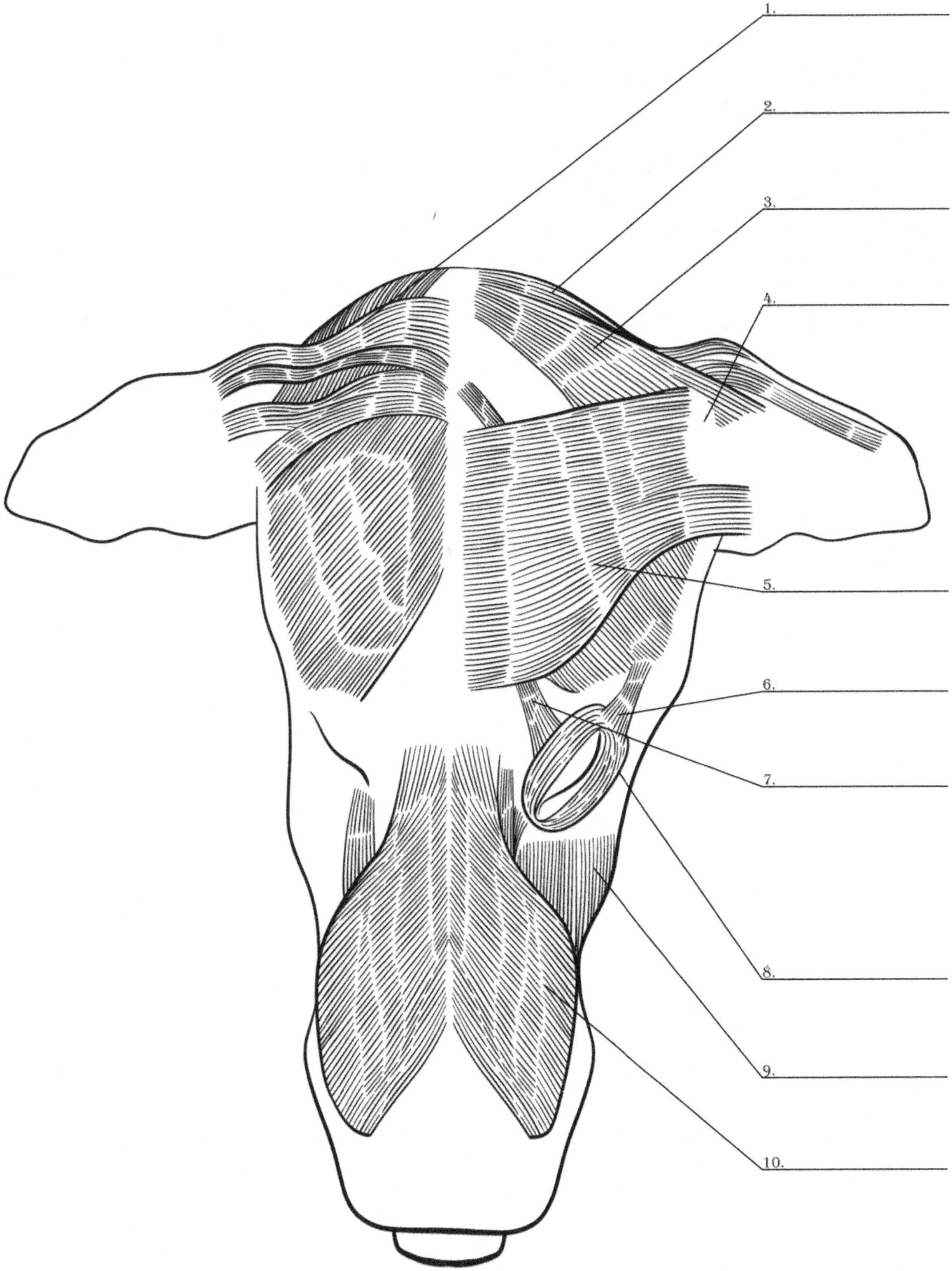

SECTION 16: THE MUSCLES OF THE HEAD DORSAL ASPECT

1. CERVICOAURICULARIS SUPERFICIALIS MUSCLE
2. CERVICOAURICULARIS PROFUNDUS MUSCLE
3. PARIETO AURICULARIS MUSCLE
4. SCUTIFORM CARTILAGE
5. FRONTOAURICULARIS & FRONTOSCUTULARIS MUSCLE
6. RETRACTOR ANGULI OCULI LATERALIS MUSCLE
7. LEVATOR ANGULI OCULI LATERALIS MUSCLE
8. ORBICULARIS OCULI MUSCLE
9. MALARIA MUSCLE
10. LEVATOR NASOLABIALIS MUSCLE

SECTION 17: THE BRAIN OF THE DOG

DORSAL VIEW

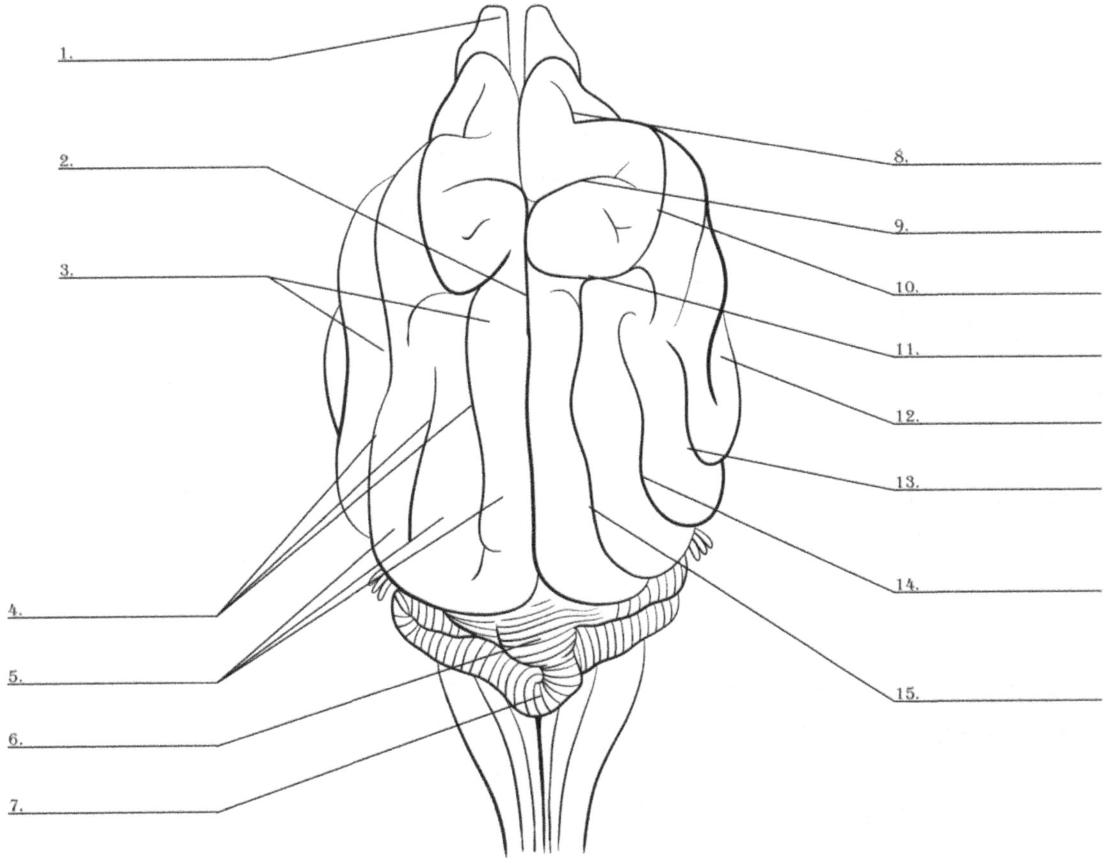

TRANSVERSE SECTION

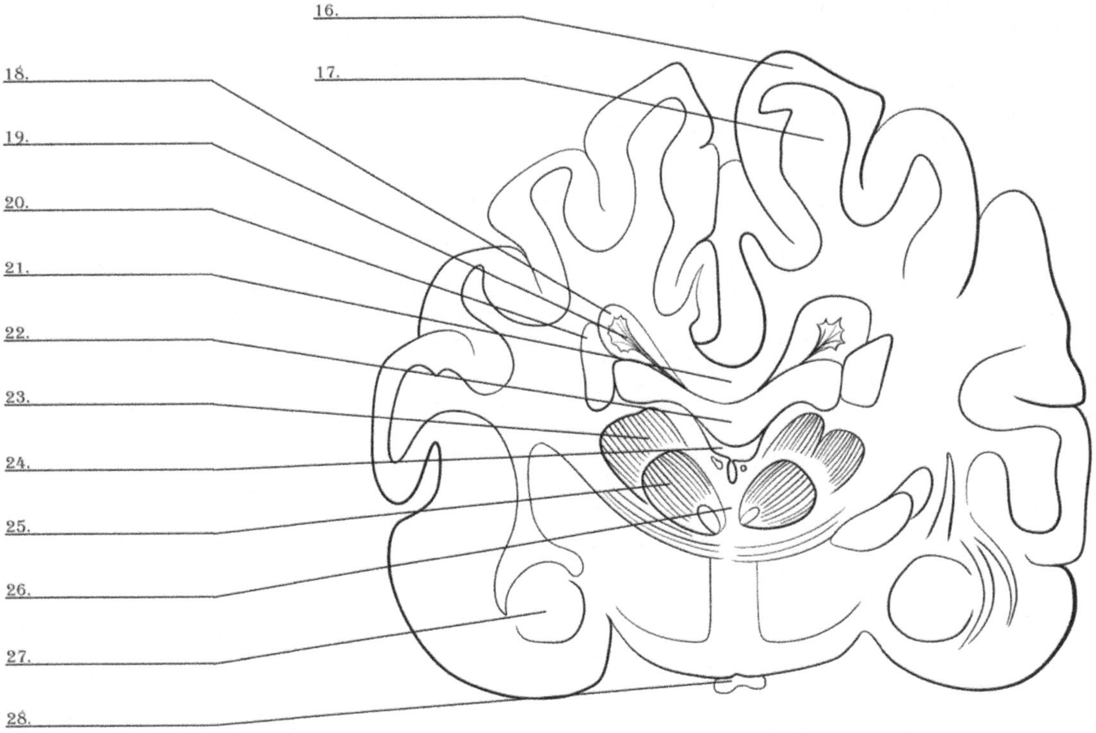

SECTION 17: THE BRAIN OF THE DOG

DORSAL VIEW
1. OLFACTORY BULB
2. LONGITUDINAL FISSURE
3. CEREBRAL HEMISPHERE
4. CEREBRAL SULCI
5. CEREBRAL GYRI
6. CEREBELLUM
7. VERMIS OF CEREBELLUM
8. PROREAN
9. CRUCIATE SULCUS
10.CORONAL SULCUS
11.ANSATE SULCUS
12.CAUDAL ECTOSYLVIAN SULCUS
13.SUPRASYLVIAN SULCUS
14.ECTOMARGINAL SULCUS
15.MARGINAL SULCUS
TRANSVERSE SECTION
16.CEREBRAL CORTEX (GRAY SUBSTANCE)
17.MEDULLA (WHITE SUBSTANCE)
18.LATERAL VENTRICLE
19.CHOROID PLEXUS OF LATERAL VENTRICLE
20.CAUDATE NUCLEUS
21.CORPUS CALLOSUM
22.FORNIX
23.ROSTRAL AND LATERAN NUCLEUS
24.THIRD VENTRICLE
25.INTERTHALAMIC ADHESION
26.SUBTHALAMIC NUCLEUS
27.EXTERNAL CAPSULE
28.OPTIC CHIASM

SECTION 18: THE EYE OF THE DOG

ROSTRAL VIEW

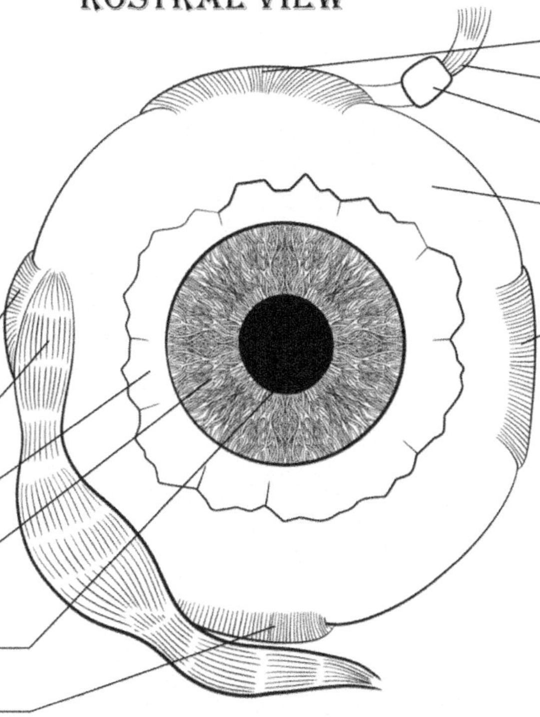

NASAL VIEW

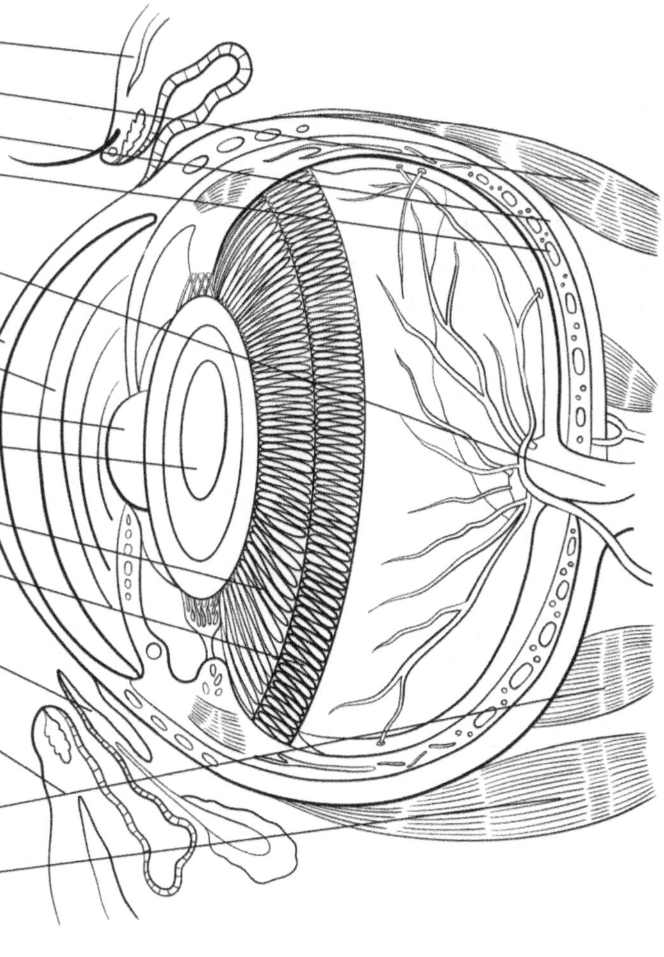

SECTION 18: THE EYE OF THE DOG

ROSTRAL VIEW
1. RECTUS DORSALIS MUSCLE
2. OBLIQUUS DORSALES MUSCLE
3. TROCHLEA
4. SCLERA
5. RECTUS MEDIUS MUSCLE
6. RECTUS LATERAL MUSCLE
7. OBLIQUUS VENTRIS MUSCLE
8. TUNICA CONJUNCTIVA OF BULB
9. IRIS
10. PUPIL
11. RECTUS VENTRIS MUSCLE

NASAL VIEW
12. SUPERIOR PALPEBRAL
13. DORSAL RECTUS MUSCLE
14. SCLERA
15. CHOROID
16. OPTIC NERVE
17. CORNEA
18. IRIS
19. PUPIL
20. LENS
21. CILIARY BODY
22. ORBICULARIS CILIARIS
23. THIRD EYELID
24. INFERIOR PALPEBRAL
25. RETRACTOR BULBI MUSCLE
26. VENTRAL RECTUS MUSCLE

SECTION 19: THE NOSE OF THE DOG

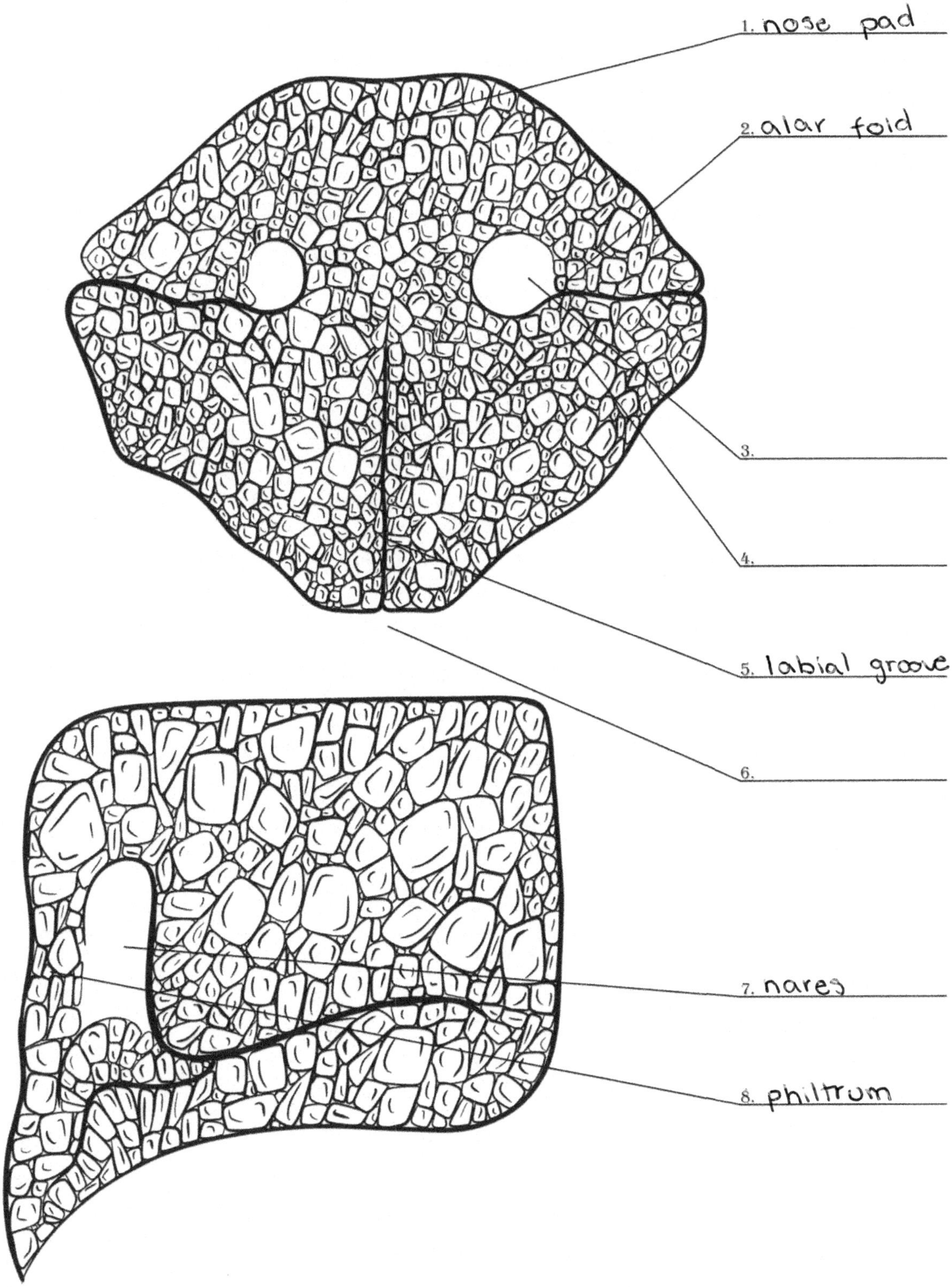

SECTION 19: THE NOSE OF THE DOG

1. NOSE PAD OR RHINARIUM
2. ALAR FOLD
3. TRUE NOSTRIL
4. FALSE NOSTRIL
5. LABIAL GROOVE
6. UPPER LIP
7. EXTERNAL NARES
8. PHILTRUM

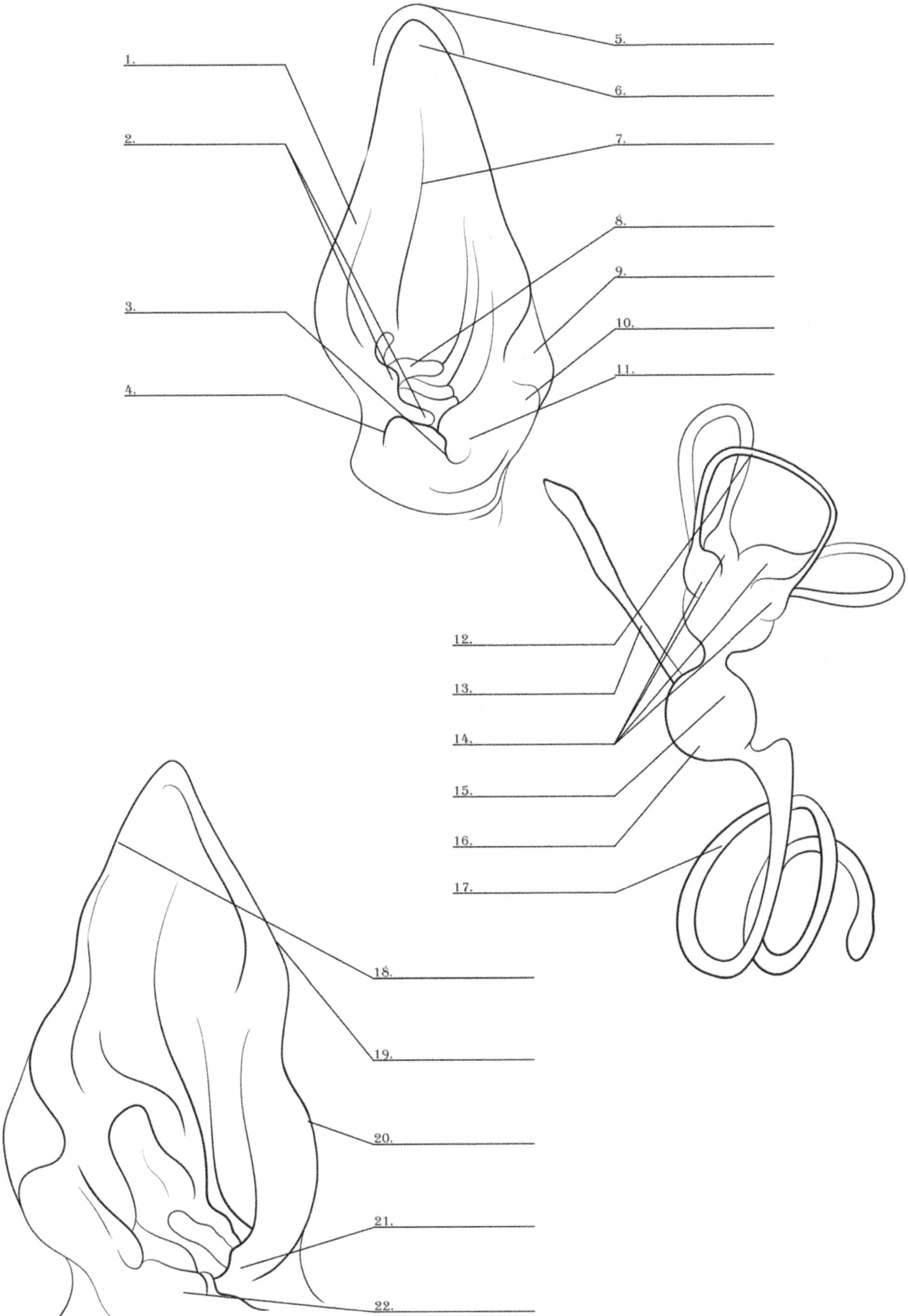
SECTION 20: THE EAR OF THE DOG
1.
2.
3.
4.
5.
6.
7.
8.
9.
10.
11.
12.
13.
14.
15.
16.
17.
18.
19.
20.
21.
22.

SECTION 20: THE EAR OF THE DOG

1. SPINA HELICIS
2. CURA HELICIS
3. INTERTRAGIC NOTCH
4. PRETRAGIC NOTCH
5. HELIX
6. APEX
7. SCAPHA
8. ANTHELIX
9. CUTANEOUS POUCH
10. CAUDA HELICIS
11. ANTITRAGUS
12. SEMICIRCULAR DUCT
13. ENDOLYMPHATIC SAC
14. MEMBRANOUS AMPULLAE
15. UTRICULUS
16. SACCULUS
17. COCHLEAR DUCT
18. LATERAL BORDER OF HELIX
19. MEDIAL BORDER OF HELIX
20. SPINE OF HELIX
21. LATERAL CRUS OF HELIX
22. TRAGUS

SECTION 21: THORACIC LIMB LATERAL ASPECT

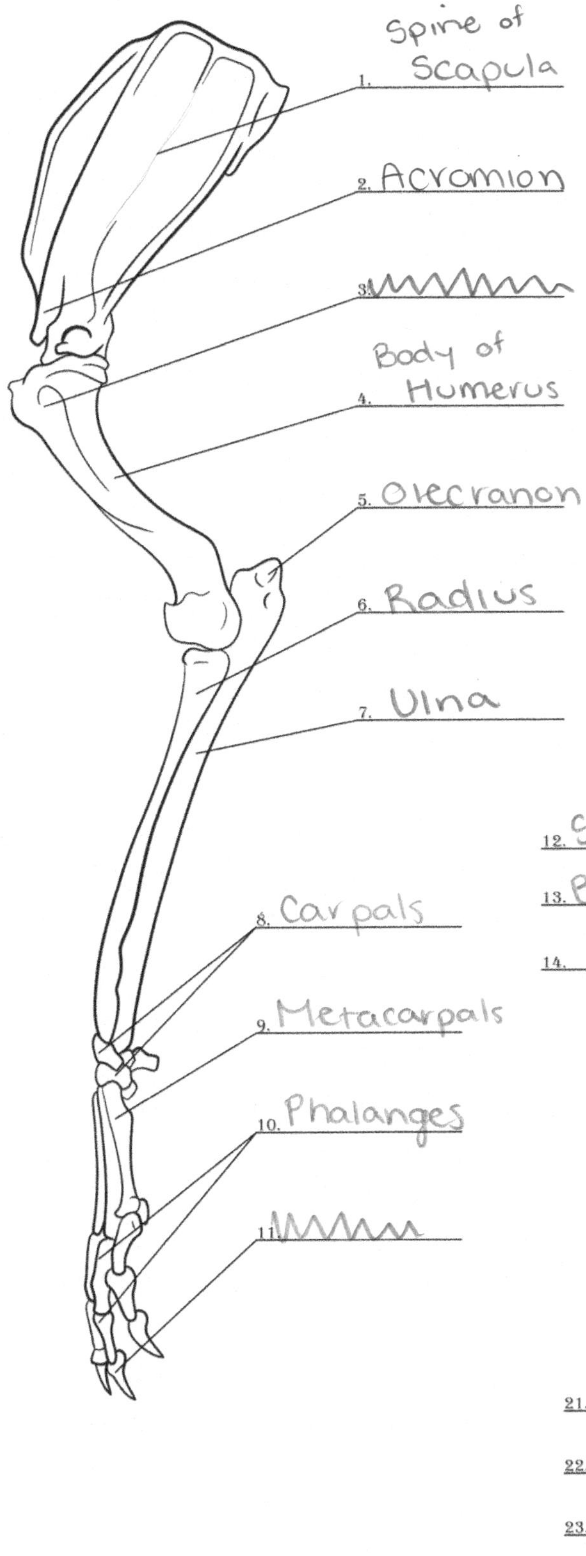

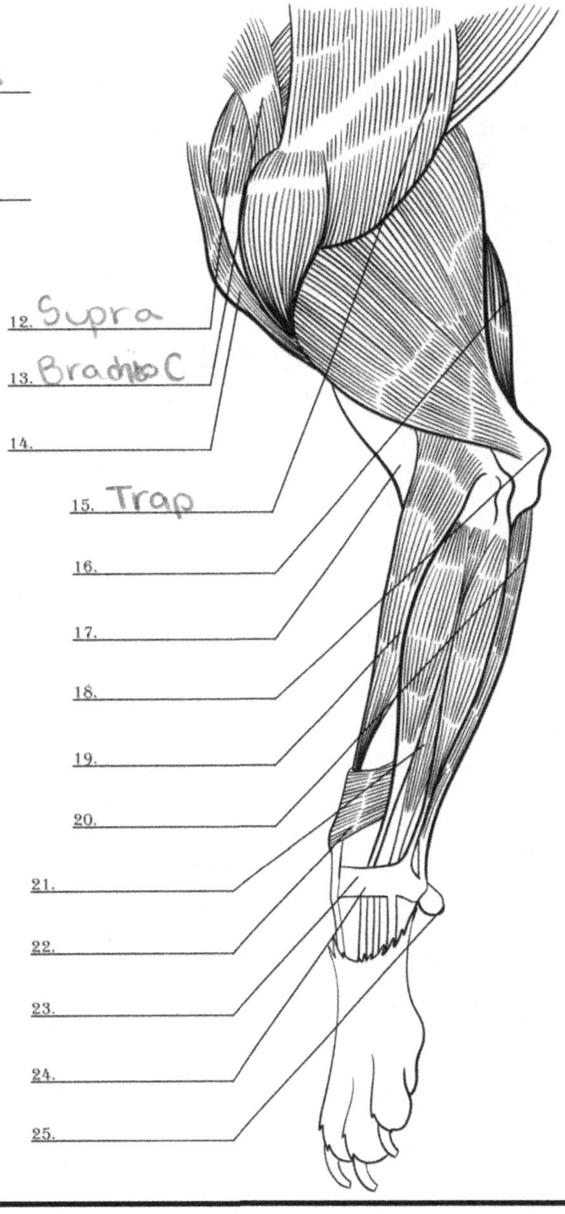

SECTION 21: THORACIC LIMB LATERAL ASPECT

1. SCAPULA
2. SPINE OF THE SHOULDER BLADE
3. MUSCULAR CONDYLE OF THE HUMERUS
4. HUMERUS
5. PROCESS OF THE ULNA
6. RADIUS
7. ULNA
8. CARPAL BONES
9. METACARPAL BONE
10. PROXIMAL AND MIDDLE PHALANGES BONES
11. CLAW BONES
12. SUPRASPINATUS MUSCLE
13. OMOTRANSVERSARIUS MUSCLE
14. BRACHIOCEPHALICUS MUSCLE
15. TRAPEZIUS MUSCLE
16. TRICEPS BRACHII MUSCLE
17. BRACHIALIS MUSCLE
18. OLECRANON
19. BRACHIORADIALIS MUSCLE
20. FLEXOR CARPI ULNARIS MUSCLE
21. EXTENSOR DIGITORUM LATERALIS MUSCLE
22. ABDUCTOR DIGITI 1ST LONGUS MUSCLE
23. EXTENSOR CARPI ULNARIS MUSCLE
24. TRANSVERSE TENDON-FIXING LIGAMENT OF CARPUS
25. CARPAL CUSHION

SECTION 22: THORACIC LIMB CRANIAL ASPECT

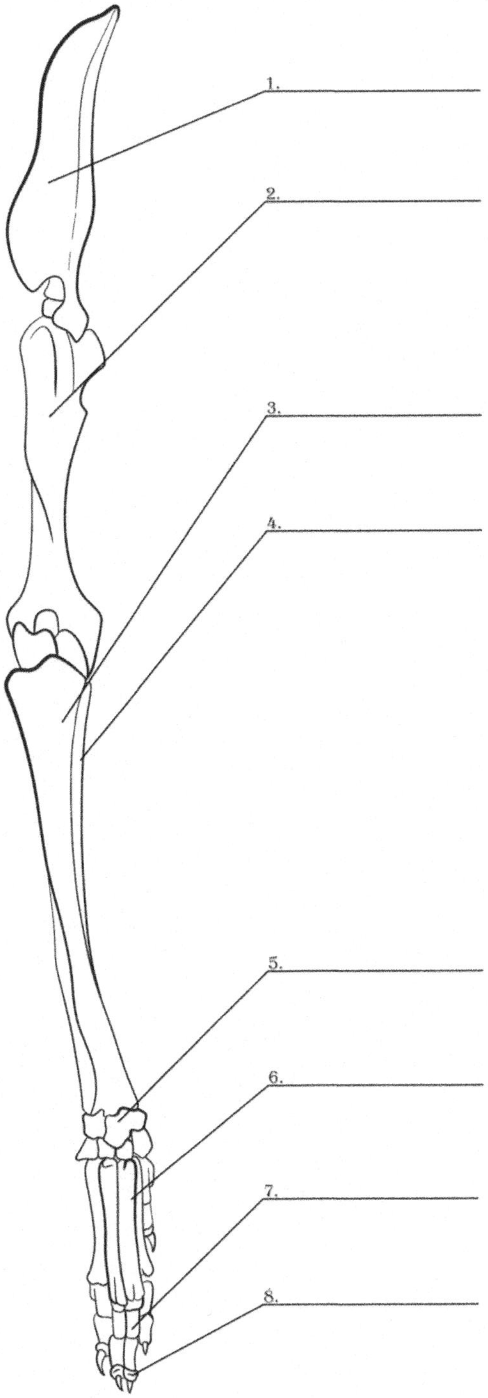

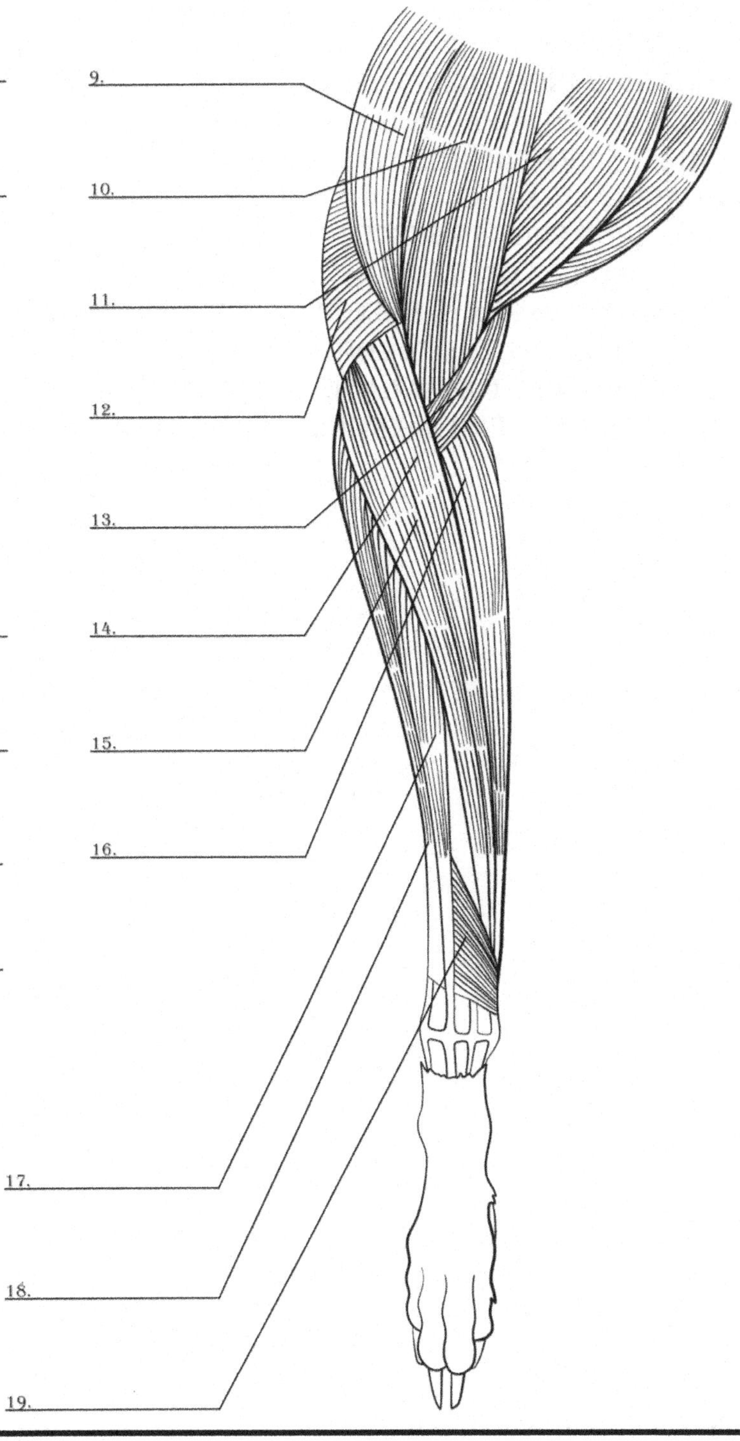

SECTION 22: THORACIC LIMB CRANIAL ASPECT

1. SCAPULA
2. HUMERUS
3. RADIUS
4. ULNA
5. CARPUS
6. METACARPUS
7. PHALANX
8. CLAW
9. DELTOIDEUS MUSCLE
10. BRACHIOCEPHALICUS MUSCLE
11. PECTORALIS SUPERFICIALIS MUSCLE
12. TRICEPS BRACHII MUSCLE
13. BRACHIALIS MUSCLE
14. BRACHIORADIALIS MUSCLE
15. EXTENSOR CARPI RADIALIS MUSCLE
16. PRONATOR TERES & FLEXOR CARPI RADIALIS MUSCLE
17. EXTENSOR DIGITORUM COMMUNIS MUSCLE
18. EXTENSOR DIGITORUM LATERALIS MUSCLE
19. TRANSVERSE TENDON-FIXING LIGAMENT OF CARPUS

SECTION 23: PELVIC LIMB LATERAL ASPECT

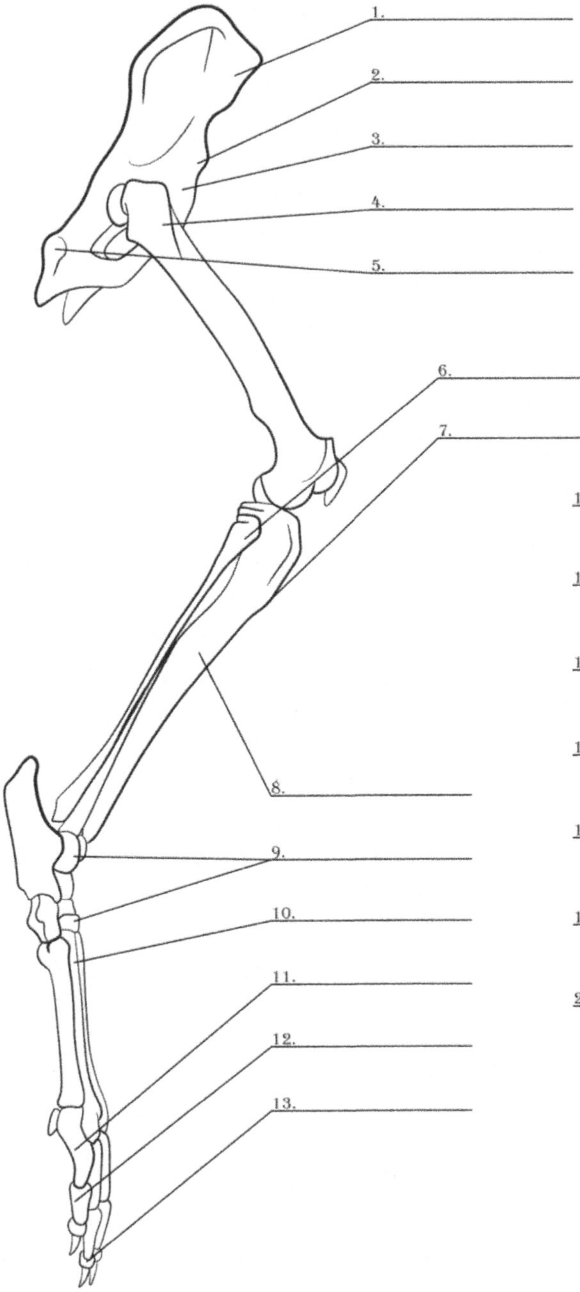

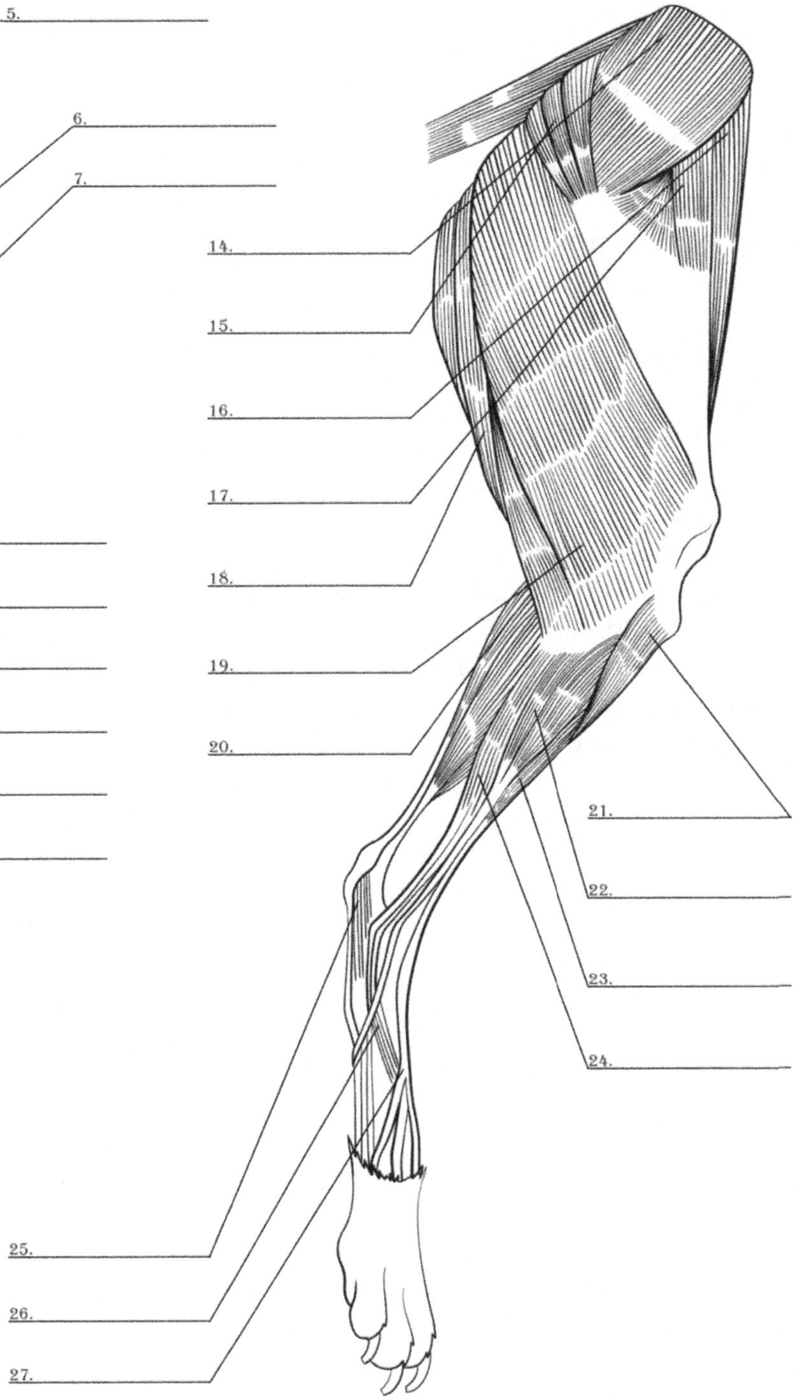

SECTION 23: PELVIC LIMB LATERAL ASPECT

1. HIPBONE
2. PUBIC BONE
3. PELVIS
4. FEMUR
5. ISCHIUM
6. FIBULA
7. TIBIAL CREST
8. TIBIA
9. TARSAL BONE
10. METATARSAL BONE
11. MIDDLE PHALANGES
12. PROXIMAL PHALANGES
13. CLAW BONE
14. GLUTEUS MEDIUS MUSCLE
15. GLUTEUS SUPERFICIALIS MUSCLE
16. SARTORIUS MUSCLE
17. TENSOR FASCIAE LATAE MUSCLE
18. SEMITENDINOSUS MUSCLE
19. BICEPS FEMORIS MUSCLE
20. TRICEPS SURAE MUSCLE
21. TIBIALIS CRANIALIS MUSCLE
22. PERONEUS LONGUS MUSCLE
23. EXTENSOR DIGITORUM LONGUS MUSCLE
24. FLEXOR HALLUCIS LONGUS MUSCLE
25. FLEXOR DIGITORUM SUPERFICIALIS MUSCLE
26. EXTENSOR DIGITORUM BREVIS MUSCLE
27. EXTENSOR DIGITORUM LATERALIS MUSCLE

SECTION 24: PELVIC LIMB CAUDAL ASPECT

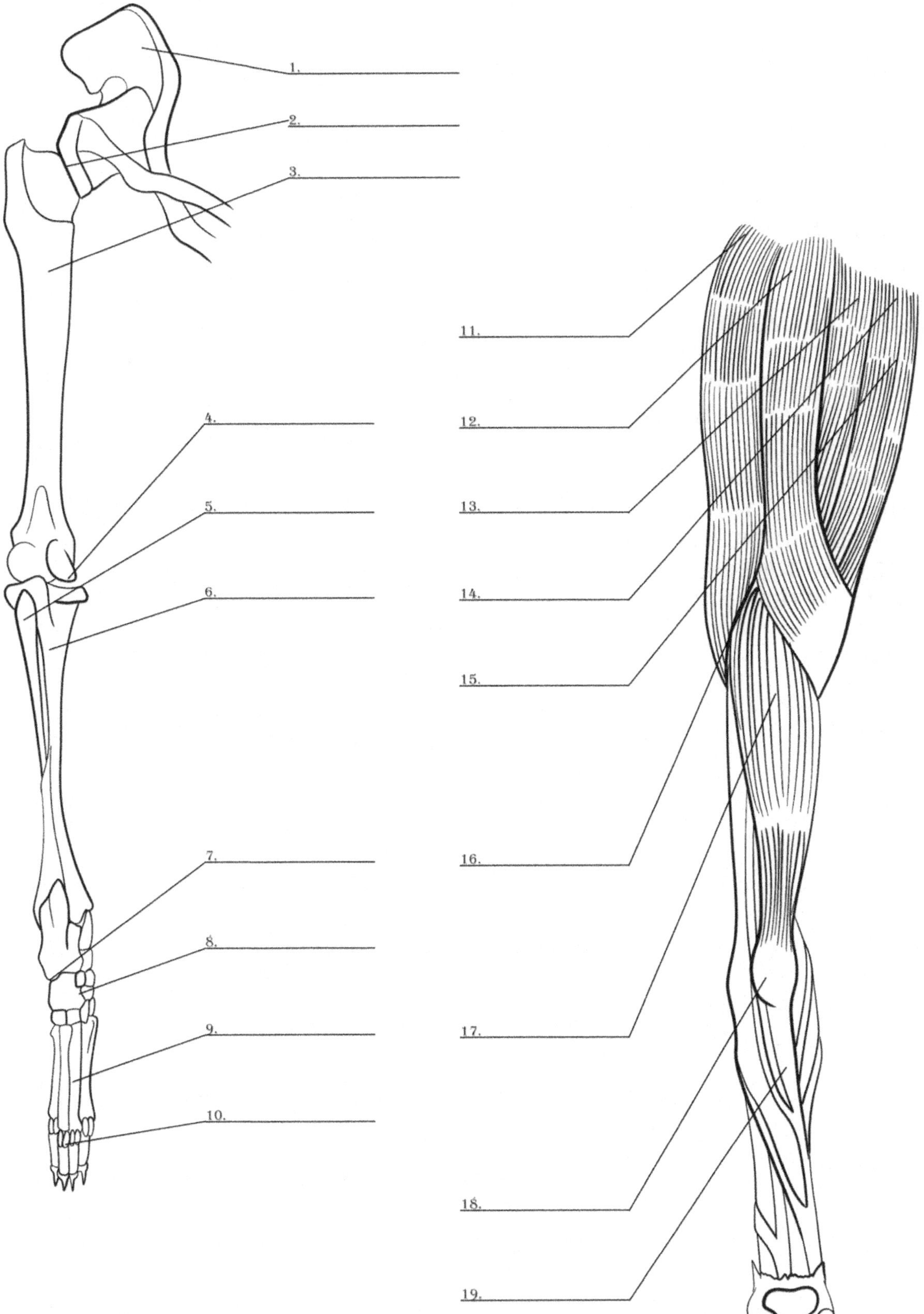

SECTION 24: PELVIC LIMB CRANIAL ASPECT

1. PELVIS
2. HIP JOINT
3. FEMUR
4. STIFLE JOINT
5. FIBULA
6. TIBIA
7. TARSAL JOINT
8. TARSUS
9. METATARSUS
10. PHALANGEAL JOINTS
11. BICEPS FEMORIS MUSCLE
12. SEMITENDINOSUS MUSCLE
13. SEMIMEMBRANOSUS MUSCLE
14. GRACILIS MUSCLE
15. SARTORIUS MUSCLE
16. ISCHIAL GROOVE
17. TRICEPS SURAE MUSCLE
18. CALQENEAU TUBEROSITY
19. TENUOUS OF THE DIGITAL FLEXORS

SECTION 25: THE PAW OF THE DOG 1

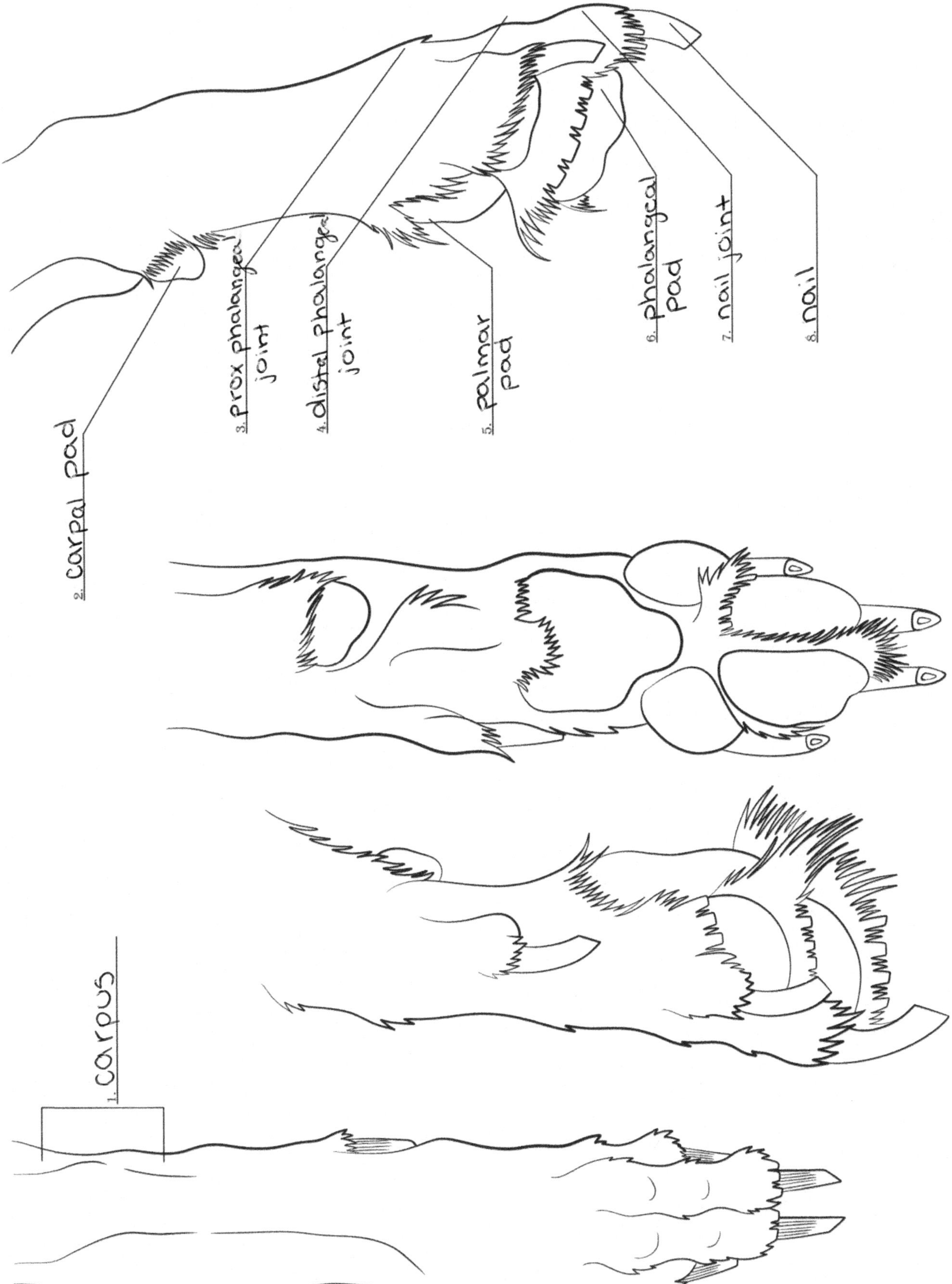

SECTION 25: THE CLAW OF THE DOG 1

1. CARPAL JOINT
2. CARPAL PAD
3. PROXIMAL PHALANGEAL JOINT
4. DISTAL PHALANGEAL JOINT
5. PALMAR PAD
6. PHALANGEAL PAD
7. CLAW JOINT
8. CLAW HORN

SECTION 26: THE PAW OF THE DOG 2

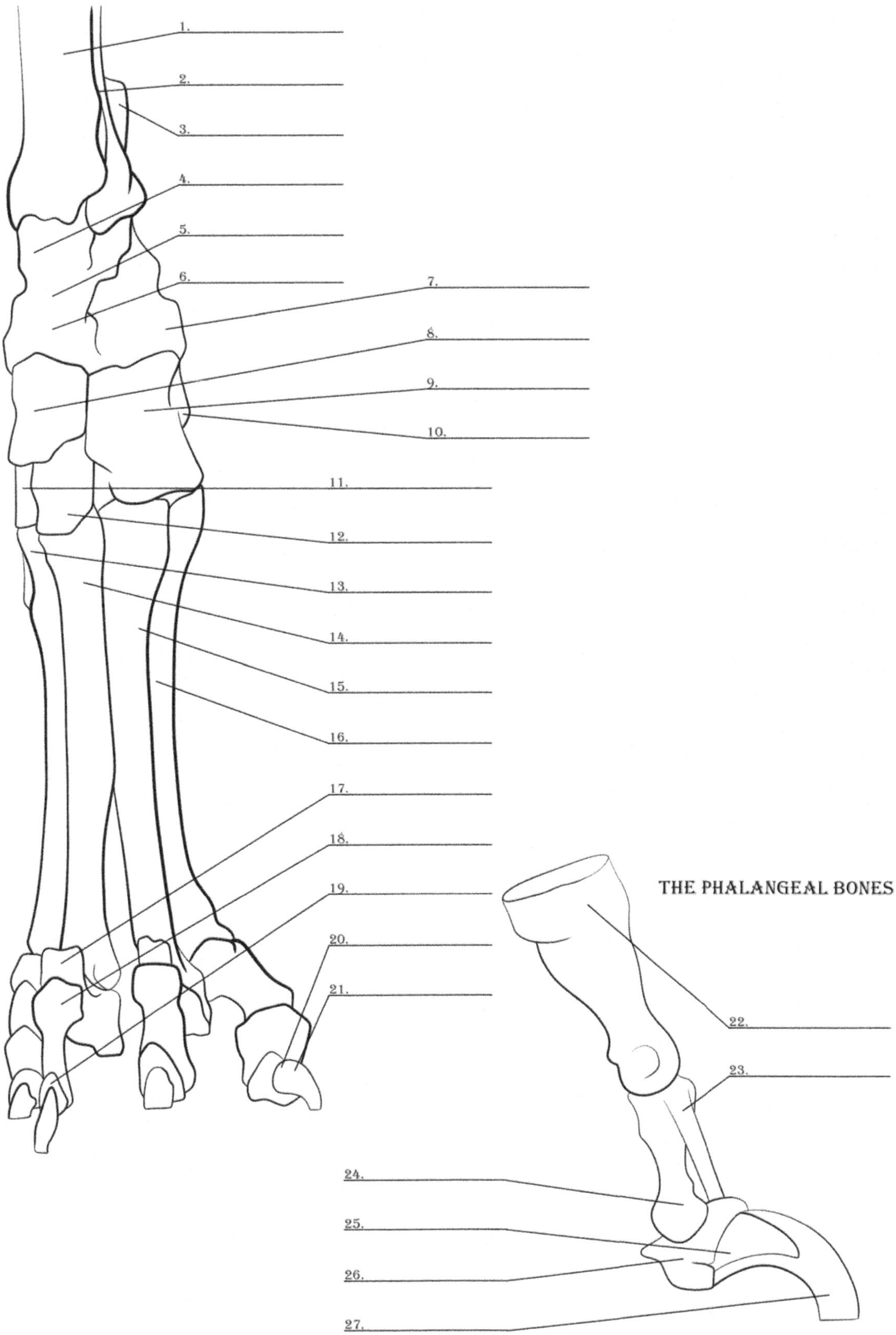

SECTION 26: THE PAW OF THE DOG 2

1. TIBIA
2. FIBULA
3. CALCANEAL TUBEROSITY
4. TALUS TROCHLEA
5. NECK
6. HEAD
7. CALCANEUS
8. CENTRAL TARSAL
9. TARSAL 4
10. SULCUS FOR MUSCLE PERONEUS LONGUS
11. TARSAL 2
12. TARSAL 3
13. METATARSAL 2
14. METATARSAL 3
15. METATARSAL 4
16. METATARSAL 5
17. PHALANX PROXIMAL
18. PHALANX MIDDLE
19. PHALANX DISTAL
20. UNGUICULAR CREST
21. UNGUICULAR PROCESS
22. PROXIMAL PHALANX
23. MIDDLE PHALANX
24. DORSAL LIGAMENT OF THE CLAW
25. GROOVE OF THE CLAW BONE
26. CLAW JOINT
27. TIP OF THE CLAW BONE

SECTION 27: THE CLAW OF THE DOG

EPIDERMIS

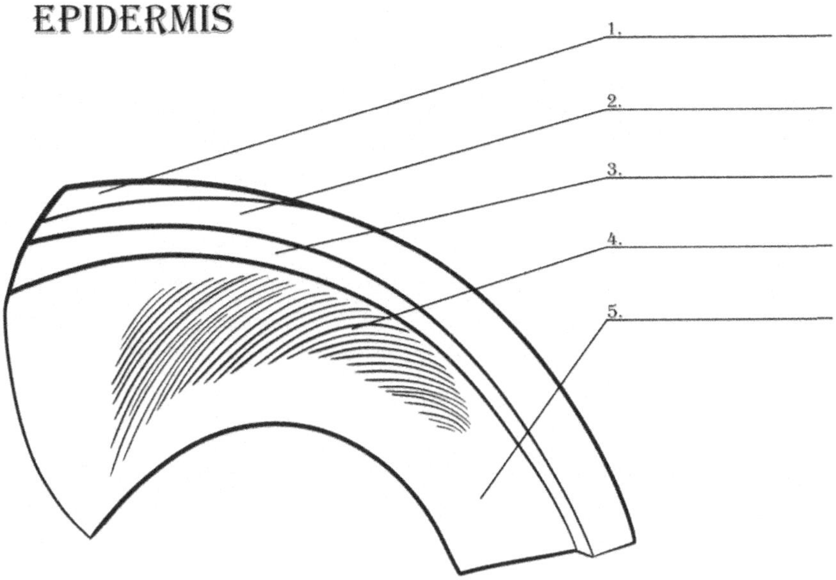

DERMIS (CORIUM)

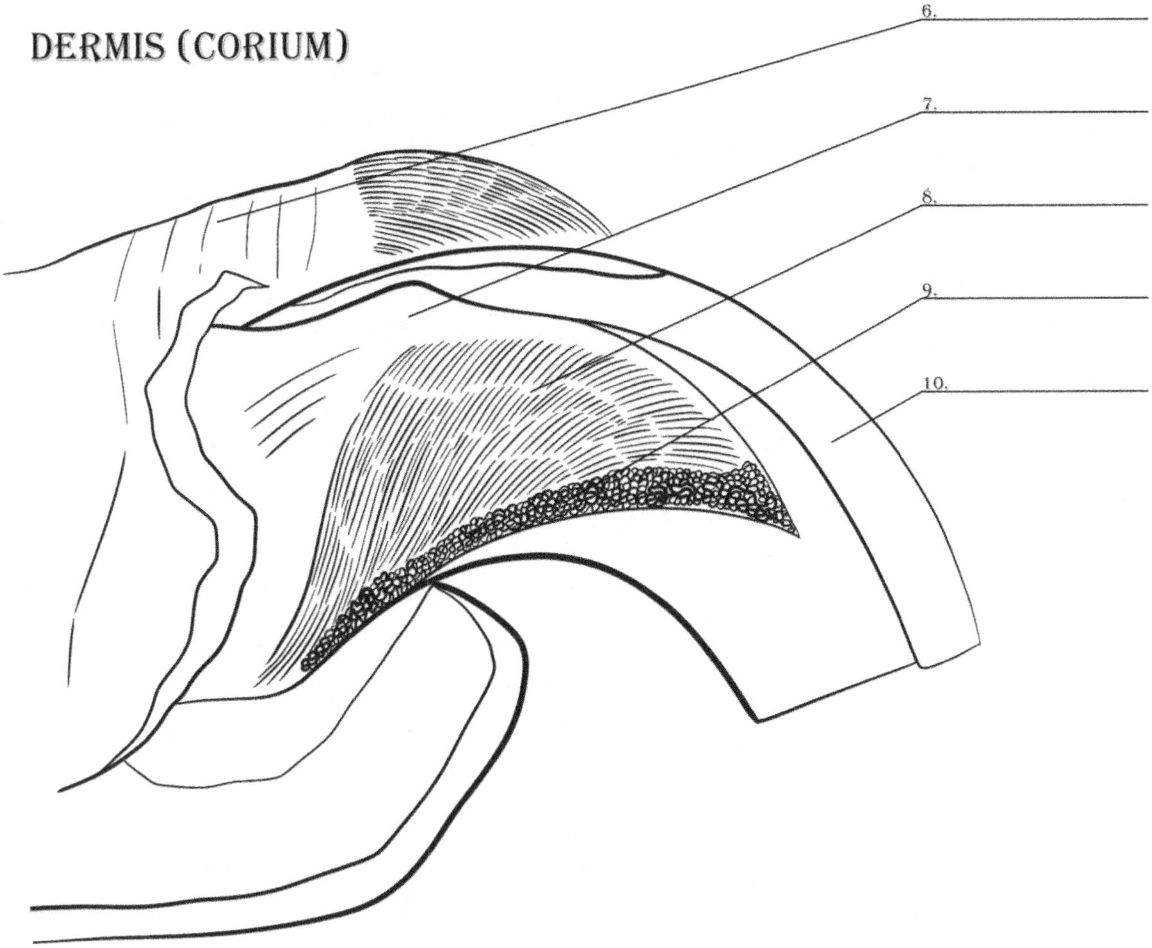

SECTION 27: THE CLAW OF THE DOG

EPIDERMIS
1. EPONYCHIUM
2. MESONYCHIUM
3. DORSAL HYPONYCHIUM
4. LATERAL HYPONYCHIUM
5. TERMINAL HYPONYCHIUM

DERMIS (CORIUM)
6. VALLUM
7. DORSUM DERAMALE
8. DERMAL LAMELLAE
9. DERMAL PAPILLAE
10. MESONYCHIUM

SECTION 28: THE HEART OF THE DOG

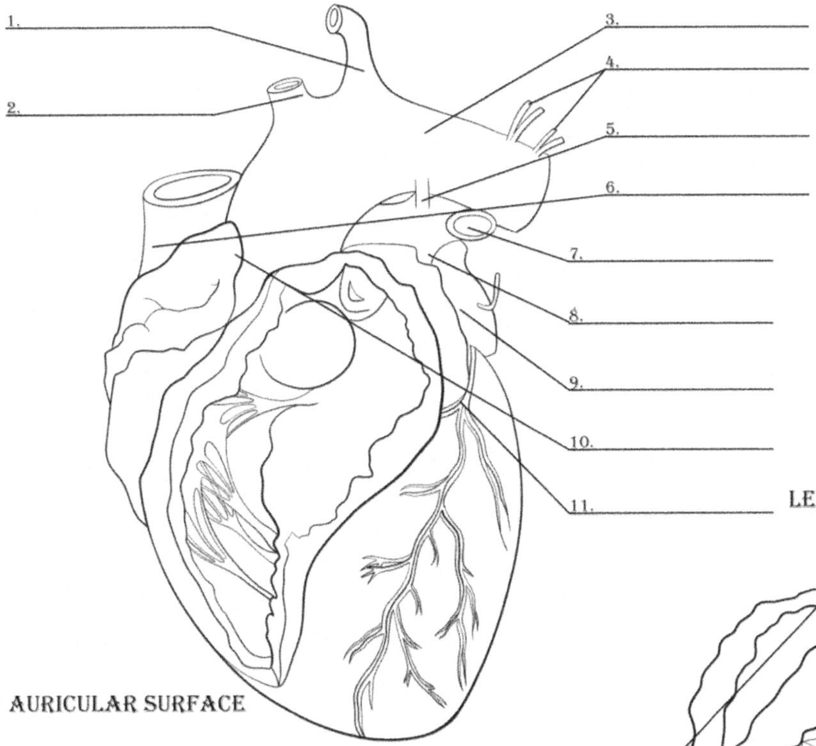

AURICULAR SURFACE

LEFT ATRIUM AND LEFT VENTRICLE

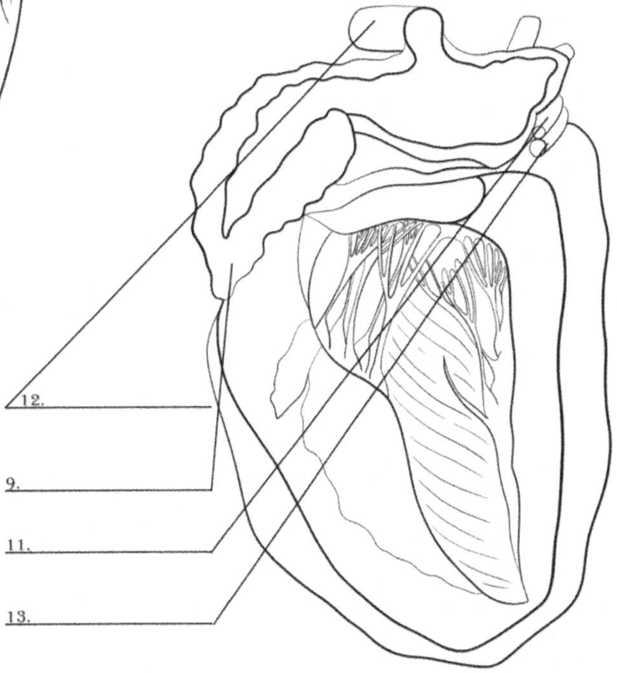

BASE OF THE HEART

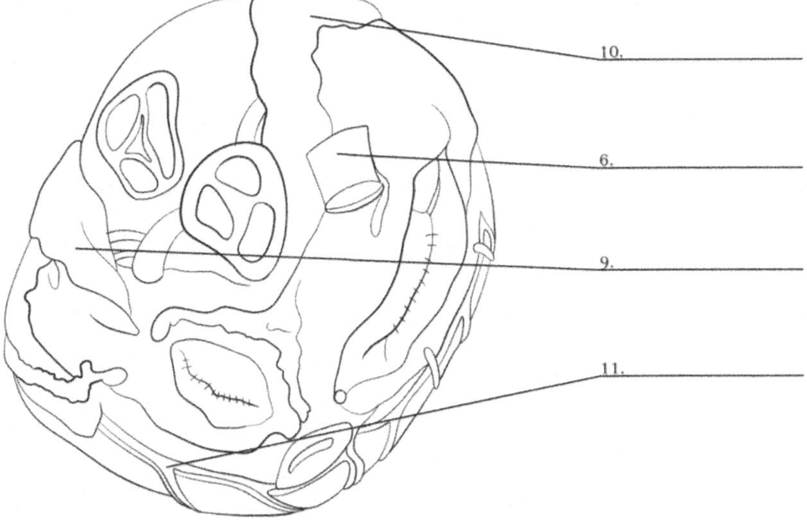

SECTION 28: THE HEART OF THE DOG

1. LEFT SUBCLAVIAN AORTA
2. BRACHIOCEPHALIC TRUNK
3. AORTA
4. INTERCOSTAL ARTERIES
5. LIGAMENT ARTERIOSUM
6. CRANIAL VENA CAVA
7. LEFT PULMONARY ARTERY
8. PULMONARY TRUNK
9. LEFT AURICLE
10. RIGHT AURICLE
11. GREAT CARDIAC VEIN
12. PULMONARY VEIN
13. CIRCUMFLEX BRANCH

SECTION 29: THE LUNGS OF THE DOG

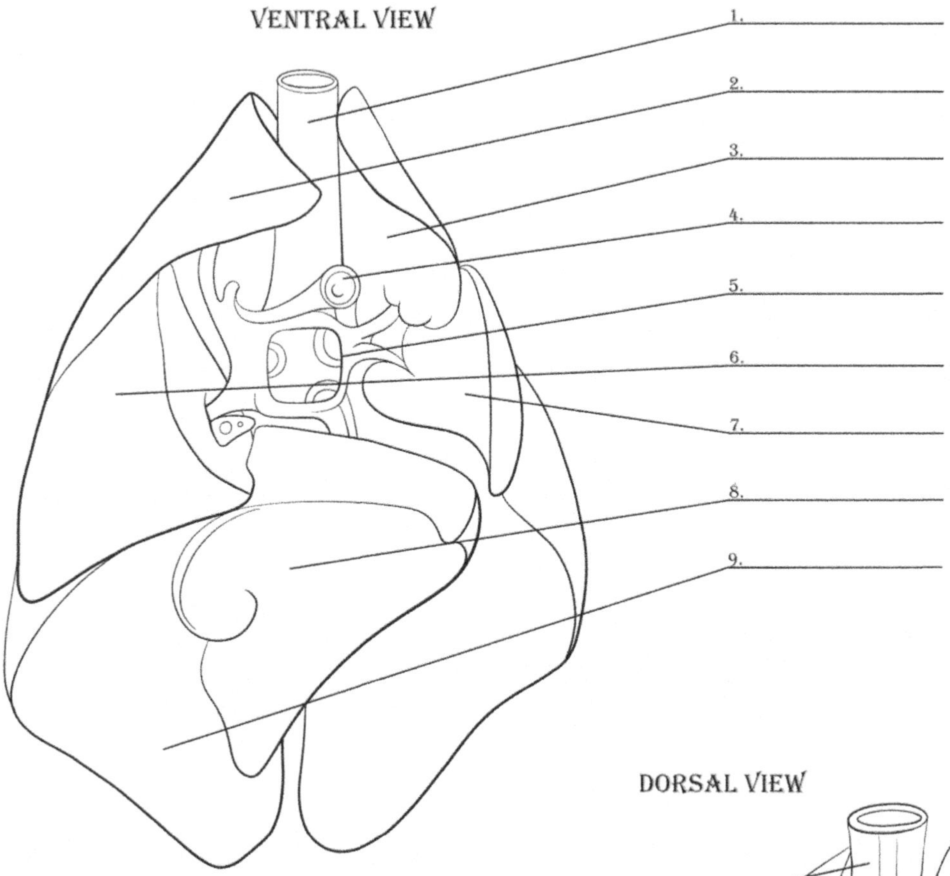

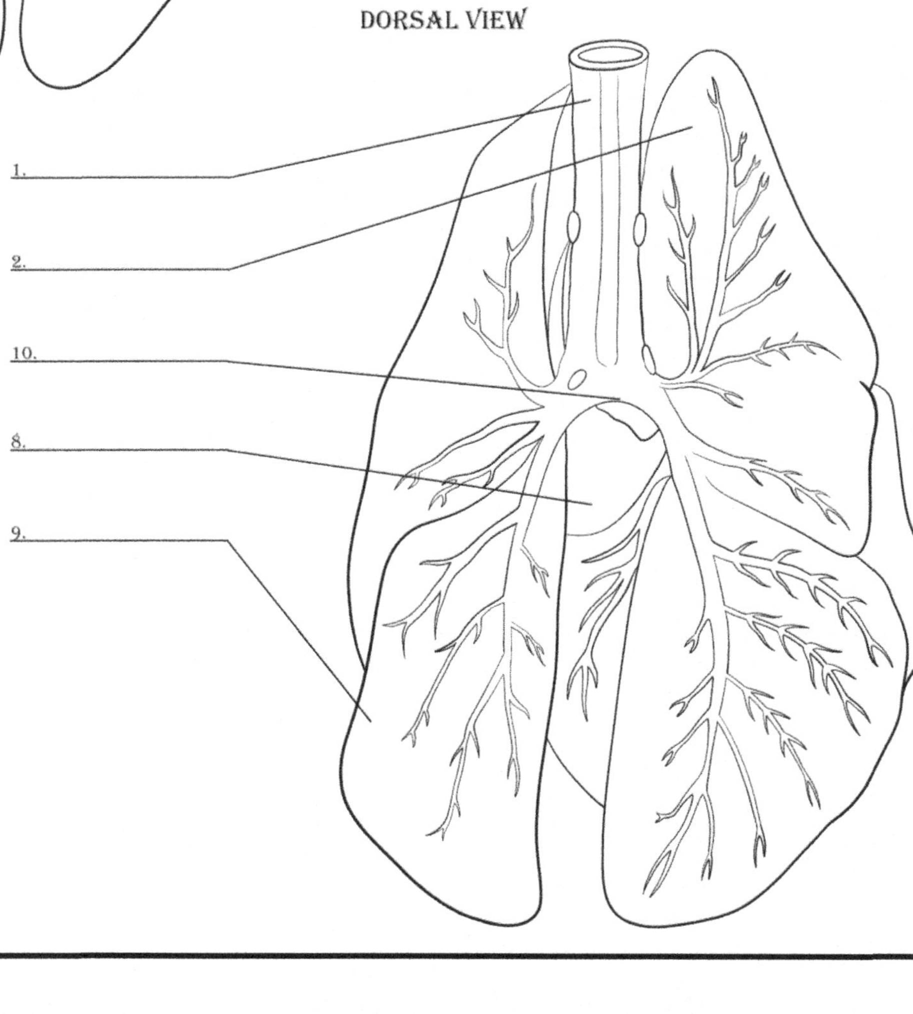

SECTION 29: THE LUNGS OF THE DOG

1. TRACHEA
2. CRANIAL LOBE
3. CRANIAL PART
4. PULMONARY TRUNK
5. PULMONARY VEINS
6. MIDDLE LOBE
7. CAUDAL PART
8. ACCESSORY LOBE
9. CAUDAL LOBE
10. BIFURCATION OF TRACHEA

SECTION 30: THE STOMACH OF THE DOG

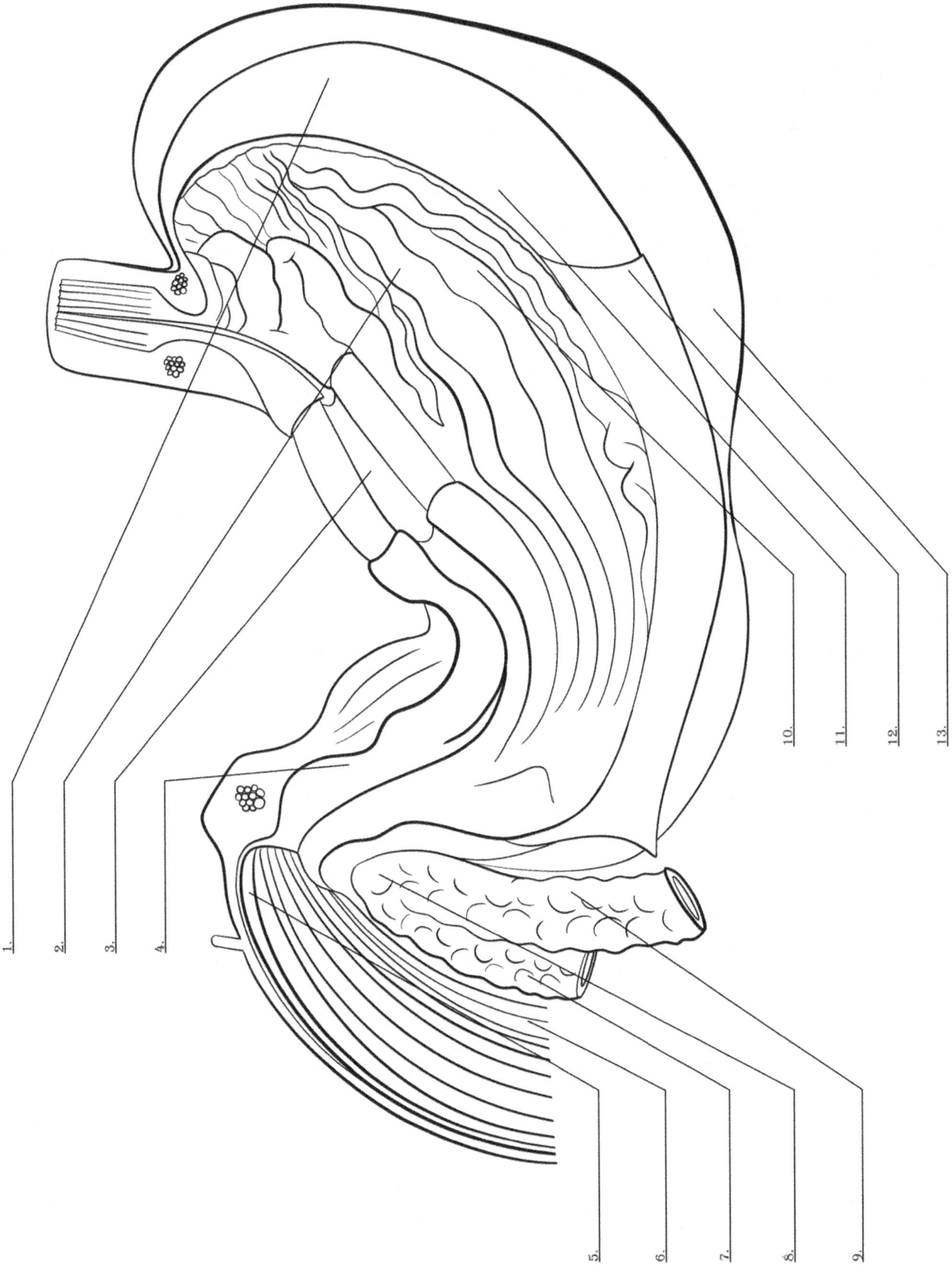

SECTION 30: THE STOMACH OF THE DOG

1. EXTRACTOR OBLIQUE FIBERS
2. MUCOUS MEMBRANE AND GASTRIC FOLDS
3. GASTRIC GROOVE
4. PYLORIC CANAL
5. CRANIAL PART OF DUODENUM
6. DESCENDING PART OF DUODENUM
7. RIGHT LOBE OF PANCREAS
8. BODY OF PANCREAS
9. LEFT LOBE OF PANCREAS
10. BODY OF STOMACH
11. LONGITUDINAL LAYER
12. CIRCULAR LAYER
13. SEROUS LAYER

SECTION 31: THE LIVER OF THE DOG

VENTRAL

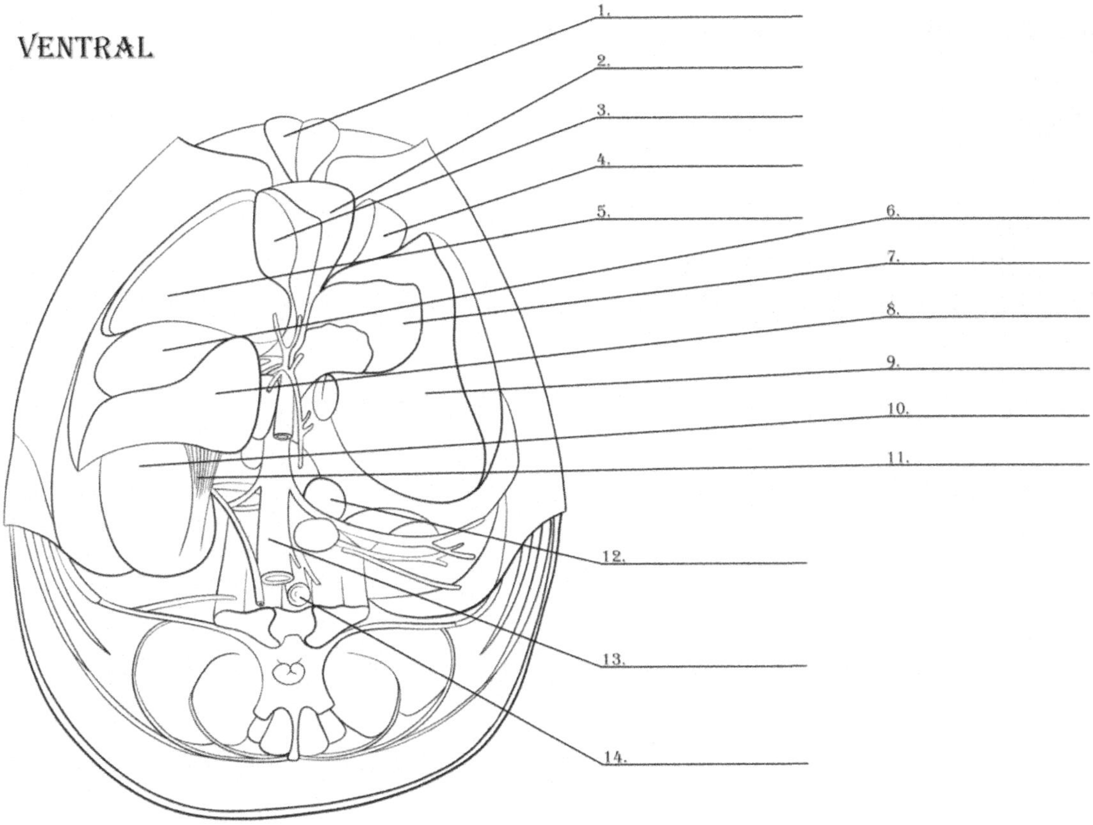

VISCLERAL SURFACE

DIAPHRAGMIC SURFACE

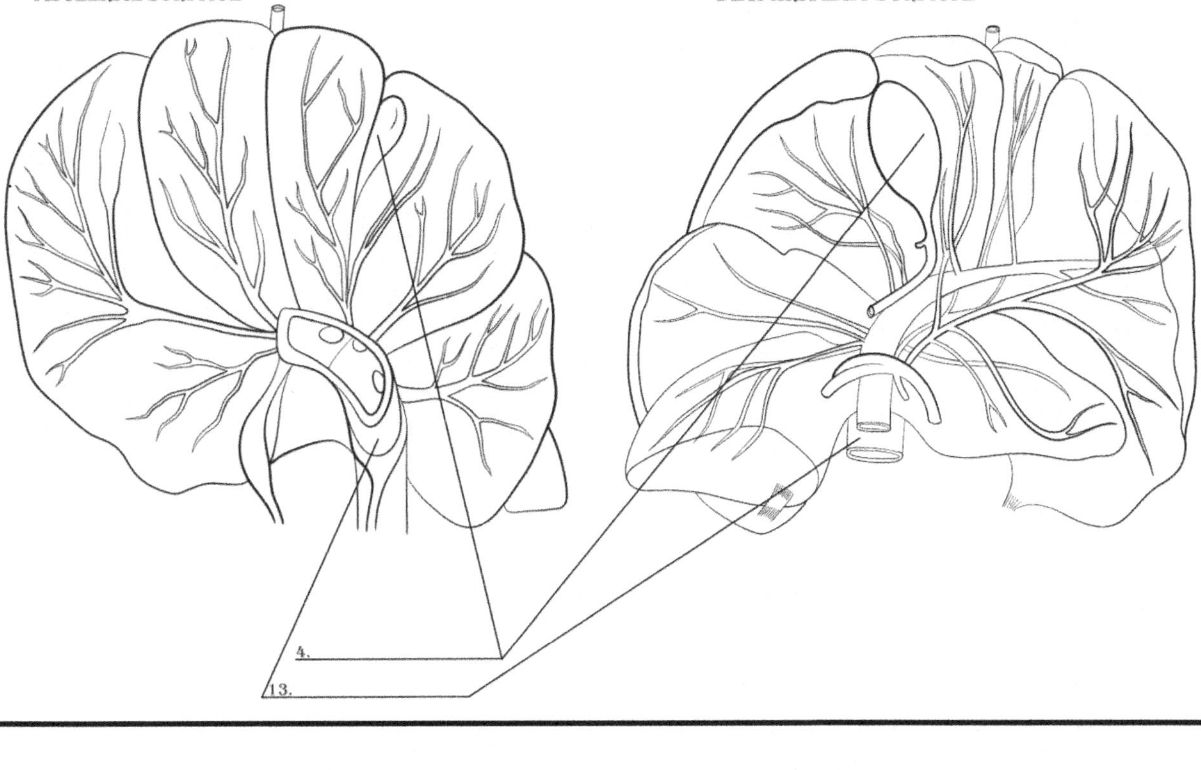

SECTION 31: THE LIVER OF THE DOG

1. FALCIFORM LIGAMENT AND ROUND LIGATURE OF LIVER
2. QUADRATE LOBE
3. GALL BLADDER
4. LEFT MEDIAL LOBE
5. RIGHT MEDIAL LOBE
6. RIGHT LATERAL LOBE
7. PAPILLARY PROCESS OF CAUDATE LOBE
8. CAUDATE PROCESS OF CAUDATE LOBE
9. LEFT LATERAL LOBE
10. RIGHT KIDNEY
11. HEPATORENAL LIGAMENT
12. ADRENAL GLAND
13. CAUDAL VENA CAVA
14. AORTA

SECTION 32: THE SPINAL CORD OF THE DOG

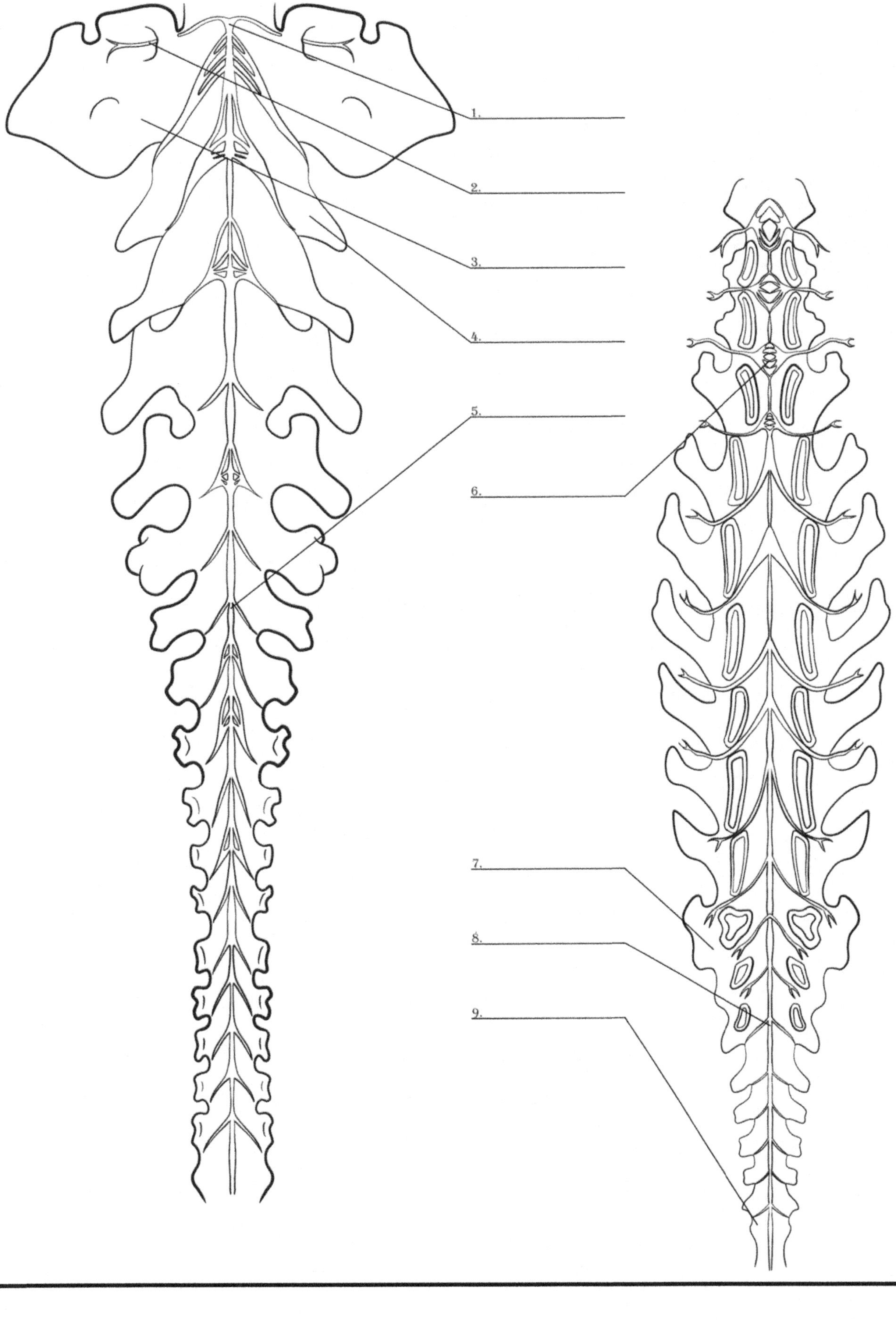

SECTION 32: THE SPINAL CORD OF THE DOG

1. CERVICAL VERTEBRAE (7)
2. NERVE
3. ATLAS
4. AXIS
5. THORACIC VERTEBRAE (13)
6. LUMBAR VERTEBRAE (7)
7. SACRUM (3)
8. COCCYGEAL (20-23)
9. FILUM TERMINALE

Made in the USA
Columbia, SC
06 July 2021

41472657R00039